The Future of Glycerol
New Uses of a Versatile Raw Material

RSC Green Chemistry Book Series

Series Editors

James H Clark, *Department of Chemistry, University of York, York, UK*
George A Kraus, *Department of Chemistry, Iowa State University, Iowa, USA*

Green Chemistry is one of the most important and rapidly growing concepts in modern chemistry. Through national awards and funding programmes, national and international courses, networks and conferences, and a dedicated journal, Green Chemistry is now widely recognised as being important in all of the chemical sciences and technologies, and in industry as well as in education and research. The RSC Green Chemistry book series is a timely and unique venture aimed at providing high level research books at the cutting edge of Green Chemistry.

Titles in the Series:

The Future of Glycerol: New Uses of a Versatile Raw Material

By Mario Pagliaro, *CNR, Institute of Nanostructured Materials and Institute for Scientific Methodology, Palermo, Italy and* Michele Rossi, *Department of Inorganic Chemistry, University of Milan, Milan, Italy*

Visit our website on www.rsc.org/books

For further information please contact:
Sales and Customer Care, Royal Society of Chemistry, Thomas Graham House, Science Park, Milton Road, Cambridge, CB4 0WF, UK
Telephone: +44 (0)1223 432360, Fax: +44 (0)1223 426017,
Email: sales@rsc.org

The Future of Glycerol
New Uses of a Versatile Raw Material

Mario Pagliaro
CNR, Institute of Nanostructured Materials and Institute for Scientific Methodology, Palermo, Italy

Michele Rossi
Department of Inorganic Chemistry, University of Milan, Milan, Italy

RSCPublishing

ISBN: 978-0-85404-124-4

A catalogue record for this book is available from the British Library

Published by The Royal Society of Chemistry,
Thomas Graham House, Science Park, Milton Road,
Cambridge CB4 0WF, UK

Registered Charity Number 207890

For further information see our web site at www.rsc.org

Dedication

This book is dedicated to Francesco and Davide Pagliaro: May they one day enjoy the pleasure of writing.

Preface

Whoever is right—Tad Patzek in pointing out that massive biofuel production will cause ecological devastation, or Shell in claiming that its new fields of *Jathropa* trees will not impact food production—it is impossible to ignore the immense quantities of glycerol resulting from biodiesel manufacture. The volumes of glycerol remaining unsold up to mid-2005 certainly paint a tragic picture of wasted energy and material resources, to say nothing of human intelligence and effort, brought about by the lack of suitable conversion processes for what is the oldest organic molecule known to man. We have even witnessed biodiesel producers experimenting by adding glycerol to animal feed, or spraying it on dirt roads to keep the dust down—and even using it as landfill, as though glycerol were spent nuclear fuel!

Following three or four years of intense research activity, chemical ingenuity worldwide has opened up a number of practical avenues for converting glycerol into value-added products. Many of these are potentially large volume outlets, and may incidentally go some way towards improving the tarnished public image of chemistry. In reviewing and commenting on these achievements this book aims to remind chemical industry professionals, both managers and technologists, of the enormous potential of glycerol as a versatile feedstock for the production of a whole range of chemicals, polymers and fuels. In the ten chapters which follow, readers will find a thorough discussion of new uses for glycerol as a raw material, many of which are already having their impact worldwide.

For example, during the devastation brought about by hurricane Katrina in 2005 the New Orleans petrochemical refineries were shut down, interrupting the supply of a variety of chemicals, including

RSC Green Chemistry Book Series
The Future of Glycerol: New Uses of a Versatile Raw Material
By Mario Pagliaro and Michele Rossi
© Mario Pagliaro and Michele Rossi 2008

ethylene and propylene glycols. This led a global manufacturer of cement additives to replace these glycols with crude glycerol from biodiesel refineries (Chapter 3). Again, in 2007 Solvay started its retrofitted plant in Tauvaux, France, where instead of making glycerol from epichlorohydrin as had been done for decades, it began to produce this epoxy resin precursor using glycerol supplied by a French biodiesel producer (Chapter 4). Thirdly, by the time this book is published the world's first plant for the manufacture of propylene glycol from glycerol will be in full production in Atlanta, USA (Chapter 5).

This monograph tells a chemical success story, the conversion of glycerol into value-added products, and it identifies the factors which have brought this about. Whether as a solvent, antifreeze or detergent, or as a monomer for textiles or drugs, new catalytic conversions have been developed for glycerol in the synthesis of products with applications ranging from everyday household items to the manufacture of fine chemicals.

Readers will see the ways in which a number of practical limitations posed by the chemistry of glycerol, such as the low selectivity of traditional catalytic conversions, have been solved by a better understanding of its fundamental chemistry and the application of catalysis technology. In addition they will find in Chapter 10 a discussion of the sustainability issues associated with bioglycerol production. The authors are convinced that chemists and chemical engineers must be in a position to present the "triple bottom line" dimensions—societal, environmental and economic—to the community, the media and to the business world. Indeed, it is an undue emphasis on sustainability on its own which has often led to controversy such as that mentioned above. It has to be accepted that both politicians and ordinary citizens are *interested* in the arguments for biodiesel and glycerol production, and in turn they want to be reassured that refineries are necessary and environmentally sound. To paraphrase Ozin, emerging biomass-based companies and research centers need young, wise, educated scientists capable of crossing the boundaries between fields and who can explain simply the advantages and problems.

Chemical research on glycerol has shown that, given a strong economic incentive, chemists can rapidly devise a whole set of upgraded processes for biorefineries, and that the integration of these in producing energy and chemicals is not just a romantic dream promoted by green-minded scientists, but an inescapable reality.

Will the biorefinery of the future make use of other platform chemicals apart from glycerol? We have no doubt about it. In 2007, one of us had the good fortune to review University of Peking's Yuan Kou's paper on direct production of fuels from wood lignin. As we write, this seminal work is undergoing further peer-review after in-depth evaluation by editors and referees of *Science* and *Angewandte Chemie*. Yet, besides known problems with scientific publishing, enormous volumes of lignin, a by-product of cellulose manufacture, are burned in power stations, a low quality outlet parallel to that sought for many years for surplus glycerol. In one sense it could be claimed that the low price of oil in the 1990s ($10 to $20 a barrel) applied a dampener on chemical ingenuity for the whole decade, since many developments were put on the shelf until a day in the future when their use would become "economically viable". Today, not only has the price of oil multiplied by a factor of 10, but the concept of energy return on energy invested (EROI) shows that in the USA domestic petroleum now returns as little as 15 Joules for every Joule invested—whereas in the 1930s the figure was 100 Joules (C. Cleveland *et al.*, *Science*, **2006**, *312*, 1746). It is exactly this decreasing trend that is forcing society globally to switch from fossil to renewable fuels, until the day when cheap and abundant solar energy becomes a reality. In this evolutionary period biofuels—and biodiesel in particular—will certainly play a role, and it inevitably follows that glycerol will remain a key raw material for many years to come. For example, following the Dumesic findings (Chapter 2) we can readily envisage a time when syngas obtained in high yield from glycerol will be used to synthesize both fuels and methanol by the Fisher–Tropsch process.

Finally, it is obvious that the greatest contributions are going to be brought about by today's students, whose creativity will produce spectacular advances. To avoid this monograph rapidly becoming out of date it is our intention that it must remain a "living" book, and that readers will have access to periodic updates posted online on the RSC website.

Mario Pagliaro, Michele Rossi
Palermo and Milan

Contents

RSC Green Chemistry Book Series
The Future of Glycerol: New Uses of a Versatile Raw Material
By Mario Pagliaro and Michele Rossi
© Mario Pagliaro and Michele Rossi 2008

About the Authors

Mario Pagliaro (b. Palermo, 1969) is a research chemist and management thinker based in Palermo at Italy's CNR, where he leads a research group and the new Institute for Scientific Methodology. His research focuses on the development of functional materials for a variety of uses and operates at the boundaries of chemistry, biology and materials science. Between 1998 and 2003 he led the management educational center, Quality College del CNR, using the resulting income to equip his laboratories and establish a research group which currently collaborates with researchers in ten countries.

Mario holds a PhD in chemistry from Palermo University (1998), the topic of his thesis being the selective oxidation of carbohydrates; mentors were David Avnir in Jerusalem and Arjan de Nooy in the Netherlands. He has also studied and worked in France and Germany. In 2005 he was appointed *Maître de conférences associé* at the Montpellier Ecole Nationale Supérieure de Chimie. Between 1993 and 1994 he worked in the Netherlands, initially at the Rijks Universiteit, Leiden, and then at the TNO Food Research Institute in Zeist. In 1998 he was with Michel Vignon at the Grenoble's CNRS, and in 2001 he joined Carsten Bolm's research group at Aachen Polytechnic. Mario has co-invented a number of novel technologies, some of which have been commercialized. He is author of the management books *Scenario: Qualità* and *Lean Banking*. He is the author of three international patents and a large number of scientific papers. Since 2004 he has organized the prestigious Seminar "Marcello Carapezza".

 **Michele Rossi** (b. Milan, 1939) is a full professor of inorganic chemistry at Milan University, where he currently teaches in the Faculty of Science. He graduated in industrial chemistry at the University of Milan in 1963. In 1974 he became Professor of Inorganic Chemistry at the University of Bari and since 1988 he has held a similar position at the University of Milan. Professor Rossi's current research is in the fascinating world of nanoscience, particularly metal-based catalysis for the activation of small molecules. His research group has discovered the surprising catalytic activity of gold nanoparticles in liquid-phase oxidation of organic compounds. Professor Rossi has been engaged by the World Gold Council for the preparation of gold on carbon catalysts as the reference standard for liquid-phase oxidation. This standard catalyst is in use throughout the world among scientific and industrial researchers.

His scientific activity, the subject of over 150 scientific papers and a number of patents, began at the prestigious school of Lamberto Malatesta and Adriano Sacco, specializing in organometallic chemistry. In one remarkable study, Sacco and Rossi discovered the first example of reversible coordination of molecular nitrogen, at room temperature and pressure, which opened the route to "nitrogen fixation", the frontline inorganic chemistry of the period 1968–1978. From this research arose the now famous compound $CoHN_2(PPh_3)_3$, which has since become a standard feature of chemistry textbooks. During this period he joined Sei Otsuka's group in Osaka, where he spent one year working on low-valency metal complexes. Later he moved into research on the catalysis of fine chemicals synthesis. Applications of catalytic hydrogenation and oxidation have been the source of several scientific contributions and patents in the technology of clean processes. Professor Rossi's research group collaborates with a number of other research groups in Italy and abroad, and is a partner in the EU Auricat research project aimed at developing the industrial application of gold catalysis.

Acknowledgements

We wish to express our sincere appreciation to the academic and industrial colleagues who have provided their expertise in reviewing some of the chapters of this book. They are the leading researchers in many of the areas discussed and this work owes much of its relevance to their admirable scientific effort. We particularly acknowledge the assistance of Dr Hiroshi Kimura (in Japan), Professors Gadi Rothenberg (University of Amsterdam), Shelley D Minteer (Saint Louis University), Elio Santacesaria (University of Naples) and John Whittall (University of Manchester), and also Drs Katarina Klepacova (Procede Group), Paolo Forni (Grace Construction Products), Jean-Luc Dubois (Arkema) and Roger Lawrence (Davy Process Technology).

None of our research would have been possible without the crucial involvement of our collaborators, particularly Rosaria Ciriminna, Cristina della Pina and Giovanni Palmisano. We thank them for their creative ideas which have led to the development of alternative uses for glycerol and new selective conversion processes, entirely unknown prior to our joining forces. Annie Jacob and Merlin Fox of RSC Publishing were instrumental in producing this book. Finally, we warmly thank Don Sanders for his help in editing the text.

Mario Pagliaro, Michele Rossi

CHAPTER 1

Glycerol: Properties and Production

1.1 Properties of Glycerol

Glycerol (1,2,3-propanetriol, Figure 1.1) is a colorless, odorless, viscous liquid with a sweet taste, derived from both natural and petrochemical feedstocks. The name glycerol is derived from the Greek word for "sweet," *glykys*, and the terms glycerin, glycerine, and glycerol tend to be used interchangeably in the literature. On the other hand, the expressions glycerin or glycerine generally refer to a commercial solution of glycerol in water of which the principal component is glycerol. Crude glycerol is 70–80% pure and is often concentrated and purified prior to commercial sale to 95.5–99% purity.

Glycerol is one of the most versatile and valuable chemical substances known to man.[1] In the modern era, it was identified in 1779, by Swedish chemist Carl W Scheele, who discovered a new transparent, syrupy liquid by heating olive oil with litharge (PbO, used in lead glazes on ceramics). It is completely soluble in water and alcohols, is slightly soluble in many common solvents such as ether and dioxane, but is insoluble in hydrocarbons.

In its pure anhydrous condition, glycerol has a specific gravity of 1.261 g mL^{-1}, a melting point of 18.2 °C and a boiling point of 290 °C under normal atmospheric pressure, accompanied by decomposition. At low temperatures, glycerol may form crystals which melt at 17.9 °C. Overall, it possesses a unique combination of physical and chemical properties (Table 1.1), which are utilized in many thousands of commercial products.[2] Indeed, glycerol has over 1500 known end uses, including applications as an ingredient or processing aid in cosmetics,

RSC Green Chemistry Book Series
The Future of Glycerol: New Uses of a Versatile Raw Material
By Mario Pagliaro and Michele Rossi
© Mario Pagliaro and Michele Rossi 2008

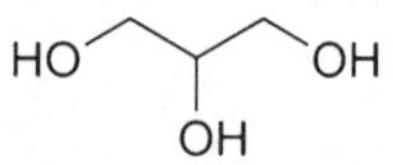

Figure 1.1   Structure of glycerol.

Table 1.1   Physicochemical properties of glycerol at 20 °C.

Chemical formula	$C_3H_5(OH)_3$
Molecular mass	$92.09382 \, g \, mol^{-1}$
Density	$1.261 \, g \, cm^{-3}$
Viscosity	1.5 Pa.s
Melting point	18.2 °C
Boiling point	290 °C
Food energy	$4.32 \, kcal \, g^{-1}$
Flash Point	160 °C (closed cup)
Surface tension	$64.00 \, mN \, m^{-1}$
Temperature coefficient	$-0.0598 \, mN \, (mK)^{-1}$

(Reproduced from Ref. 2, with permission)

toiletries, personal care products, pharmaceutical formulations and foodstuffs.[3] In addition, glycerol is highly stable under normal storage conditions, compatible with many other chemical materials, virtually non-irritating in its various uses, and has no known negative environmental effects.

Glycerol contains three hydrophilic alcoholic hydroxyl groups, which are responsible for its solubility in water and its hygroscopic nature. It is a highly flexible molecule forming both intra- and intermolecular hydrogen bonds. There are 126 possible conformers of glycerol, all of which have been characterized in a recent study using density functional theory (DFT) methods.[4] In particular, a systematic series of *ab initio* molecular orbital and density functional theory optimizations of all the possible staggered conformers of glycerol (including calculation of the Boltzmann distributions in the gas and aqueous phases) has indicated that the enthalpic and entropic contributions to the Gibbs free energy are important for an accurate determination of the conformational and energetic preferences of glycerol. In the lowest energy conformer at the CCSD–(T)/6−31 + G(d,p)//HF/6−31G(d) and CBS–QB3 levels, the hydroxyl groups form a cyclic structure with three internal hydrogen bonds. This conformer, termed gG′g,g′Gg (γγ), has the structure with the internal hydrogen bond lengths displayed in Figure 1.2 and provides the starting geometry for the mechanism of many reactions with practical applications. This conformer (*left* in Figure 1.2) is the only form of glycerol which possesses three intramolecular hydrogen bonds, a structure offering considerable energetic benefit.

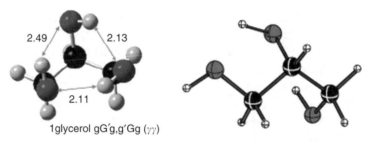

1glycerol gG'g,g'Gg ($\gamma\gamma$)

Figure 1.2 Selected bond lengths in Å for glycerol in its lowest energy conformer in the gas (*left*) and liquid (*right*) phase as determined by DFT methods. (Reproduced from Ref. 4, with permission.)

In the aqueous phase, glycerol is stabilized by a combination of intramolecular hydrogen bonds and intermolecular solvation of the hydroxyl groups. Indeed, taking into account solvation, the conformer with two intramolecular hydrogen bonds shown on the right in Figure 1.2 is the most energetically stable. This is due to the fact that in aqueous solution conformation **1** is now of higher relative energy, because all three hydroxyl groups are involved in intramolecular hydrogen bonding and are therefore unavailable to interact with the solvent. Many structures that possess intramolecular hydrogen bonding arrays still provide low energy conformations in aqueous solution, even when compared to structures without intramolecular hydrogen bonding.

In condensed phases, glycerol is characterized by a high degree of association due to hydrogen bonding. A first molecular dynamics simulation suggests that on average 95% of molecules in the liquid are connected.[5] This network is very stable and very rarely, especially at high temperature, releases a few short living (less than 0.5 ps) monomers, dimers or trimers. In the glassy state, a single hydrogen-bonded network is observed, involving 100% of the molecules present.

A highly branched network of molecules connected by hydrogen bonds exists in all phases and at all temperatures. The average number of hydrogen bonds per molecule ranges from about 2.1 in the glassy state to 1.2 in the liquid state at high temperature, with an average activation energy of 6.3 kJ mol^{-1} required to break the hydrogen bond. Crystallization, which occurs at 291 K, cannot be directly achieved from the liquid state but requires special procedures. Due to the existence of such an extended hydrogen bonded network, the viscosity and the boiling point of glycerol are unusually high. Glycerol readily forms a supercooled liquid which, by lowering the temperature undergoes at about 187 K transition to a glassy state whose nature has been the subject of a number of investigations. Remarkably, a recent single-molecule analysis has revealed a foam-like structure for glycerol at temperatures above the glass transition point (T_g, 190 K) comprising pockets of fluid isolated from one another by glass-like regions, which retain their distinct dynamics over surprisingly long timescales.[6]

The intramolecular energy is the prevailing factor in determining the average molecular structure in the condensed phases. The intermolecular hydrogen bonds, however, do not significantly stabilize energetically unfavored gas phase structures. In the gas phase in the range 300–400 K glycerol in practice exhibits only three backbone conformations, namely αα, αγ and γγ (Figure 1.3).

Intermolecular interactions stabilize slightly the αα-conformation and destabilize the γγ-conformation as the temperature decreases (Table 1.2). As the liquid approaches the freezing point about half of the molecules are present in the αα-conformation.

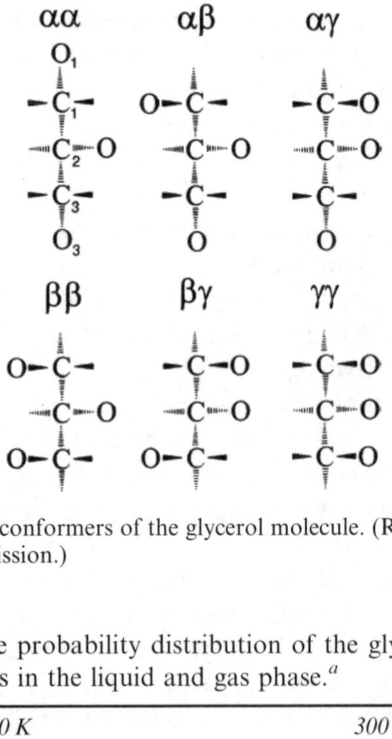

Figure 1.3   Backbone conformers of the glycerol molecule. (Reproduced from Ref. 5, with permission.)

Table 1.2   Percentage probability distribution of the glycerol backbone conformations in the liquid and gas phase.[a]

	400 K		300 K	
Conformer	Liquid	Gas	Liquid	Gas
αα	44 (3)	39.9 (5)	48 (1)	43.0 (8)
αγ + γα	45 (2)	47.0 (3)	46.0 (6)	46.9 (5)
γγ	6 (1)	11.0 (2)	4.4 (6)	9.8 (4)
αβ + βα	3.8 (9)	1.40 (1)	1.4 (4)	0.22 (1)
βγ + γβ	1.2 (5)	0.67 (1)	0.2 (1)	0.11 (1)
ββ	0.01 (1)	0.03 (1)	0	0.002 (1)

[a]Errors in the last digit are given in parenthesis. (Reproduced from Ref. 5, with permission.)

The activation energy for a conformational transition is around $20\,kJ\,mol^{-1}$, in other words it is much higher at room temperature than at freezing temperature. The high activation energy, and the fact that nearly half of the molecules must undergo at least one conformational transition to reach the $\alpha\alpha$-structure typical of the crystalline state, is responsible for the remarkable stability of supercooled glycerol. The network dynamics involved spans three distinct and increasingly longer time scales, due to vibrational motion, to neighbor exchange and to translational diffusion, respectively.

In biochemistry, glycerol plays a major role in stabilizing enzymes due to the action of polyhydric alcohol functions,[7] a fact which is generally attributed to the enhancement of the structural stability of the entire protein by a large alteration in the hydrophilic–lipophilic balance (HLB) upon clustering around the protein.[8] This results in practical implications of dramatic importance, as glycerol also protects biologicals during sol–gel entrapment in a silica-based matrix, either by formation of poly(glyceryl silicate) as sol–gel precursors[9] or by direct addition to bacteria prior to the sol–gel polycondensation.[10] This enables reproducible and efficient confinement of proteins, cells and bacteria inside hybrid bio-doped glasses. These materials display activity approaching or even exceeding those of the free biologicals, together with the high stability and robustness that characterizes sol–gel bioceramics, and accordingly are finding application over a wide field, ranging from biocatalysis to biosensing and biodiagnostics. Finally, it is worth recalling at this point that theoretical physicist David Bohm formulated the implicate order theory,[11] inspired by a simple experiment in which a drop of ink was squeezed on to a cylinder of glycerol. When the cylinder rotated (Figure 1.4), the ink diffused through the glycerol in an

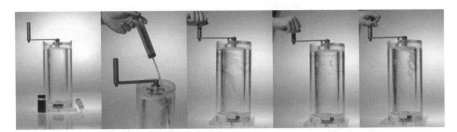

Figure 1.4 Glycerol with a high concentration of dye is inserted into the space between two concentric transparent plastic cylinders, the space between them filled with clear glycerol. The inner cylinder is then slowly and steadily rotated. As the cylinder rotates, the dye smears and appears to be diffused. Since the flow of the viscous glycerol is laminar, however, there is little actual diffusion. When the inner cylinder is rotated in the opposite direction, the dye regains the initial profile. (Reproduced from Ref. 12, with permission.)

apparently irreversible fashion; its order seemed to have disintegrated. When the inner cylinder was rotated in the opposite direction, the ink recovered its initial profile.

1.2 Traditional Commercial Applications

Traditional applications of glycerol, either directly as an additive or as a raw material, range from its use as a food, tobacco and drugs additive to the synthesis of trinitroglycerine, alkyd resins and polyurethanes (Figure 1.5).[13]

Currently, the amount of glycerol that goes annually into technical applications is around 160 000 tonnes and is expected to grow at an annual rate of 2.8%.[1] Of the glycerol market, pharmaceuticals, toothpaste and cosmetics account for around 28%, tobacco 15%, foodstuffs 13% and the manufacture of urethanes 11%, the remainder being used in the manufacture of lacquers, varnishes, inks, adhesives, synthetic plastics, regenerated cellulose, explosives and other miscellaneous industrial uses. Glycerol is also increasingly used as a substitute for propylene glycol.

As one of the major raw materials for the manufacture of polyols for flexible foams, and to a lesser extent rigid polyurethane foams, glycerol is the initiator to which propylene oxide and ethylene oxide is added. Glycerol is widely used in alkyd resins and regenerated cellulose as a

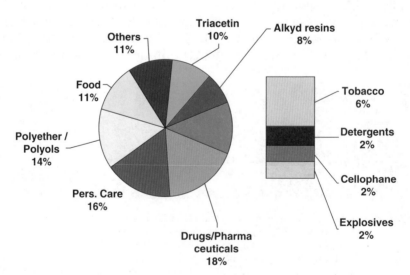

Figure 1.5 Market for glycerol (volumes and industrial uses). (Source: Novaol, May 2002.)

softener and plasticizer to impart flexibility, pliability and toughness in surface coatings and paints.

Most of the glycerol marketed today meets the stringent requirements of the United States Pharmacopeia (USP) and the Food Chemicals Codex (FCC). However, technical grades of glycerol that are not certified as USP or FCC quality are also found. Also available is Food Grade Kosher glycerol, which has been prepared and maintained in compliance with the customs of the Jewish religion.[14]

The primary function of glycerol in many cases is as a humectant, a substance for retaining moisture and in turn giving softness. Glycerol draws water from its surroundings and the heat produced by the absorption makes it feel warm. Due to this property, glycerol is added to adhesives and glues to keep them from drying too rapidly. Many specialized lubrication problems have been solved by using glycerol or glycerol mixtures. Many thousand tonnes of glycerol are used each year to plasticize a variety of materials such as sheeting and gaskets. The flexibility and toughness of regenerated cellulose films, meat casings and special quality papers can be attributed to the presence of glycerol. It also acts as a solvent, sweetener, and preservative in food and beverages, and as a carrier and emollient in cosmetics. The effectiveness of glycerol as a plasticizer and lubricant gives it wide applicability, particularly in food processing, because it is nontoxic. Glycerol is also used in alkyd resin manufacture to impart flexibility. Alkyd resins are used as binders in products such as paints and inks, where brittleness is undesirable. Glycerol is used in specialist lubricants where oxidation stability is required, for example in air compressors. In all applications, whether as a reactant or as an additive, the non-toxicity and overall safety of glycerol is a significant benefit.

Due to the rapid decline in its price, glycerol is rapidly substituting other polyols which are used on a large scale as sugar-free sweeteners. Polyols are used mostly in confectionery, food, oral care, pharmaceutical, and industrial applications. Some characteristics of polyols are reduced calories, a pleasant sweetness, the ability to retain moisture, and improved processing. The most widely used polyols are sorbitol, mannitol, and maltitol. Sorbitol is facing particularly stiff competition from glycerol. Glycerol contains approximately 27 calories per teaspoonful and is 60% as sweet as sucrose; it has about the same food energy as table sugar. However, it does not raise blood sugar levels, nor does it feed the bacteria that cause plaque and dental cavities. As a food additive, glycerol is coded E422. Baked goods lose their appeal when they become dry and hard during storage. Being hygroscopic, glycerol reduces water loss and prolongs shelf life.

Glycerol is used in medical and pharmaceutical preparations, mainly as a means of improving smoothness, providing lubrication and as a humectant, that is as a hygroscopic substance which keeps the preparation moist. Glycerol helps to maintain texture and adds humectancy, controls water activity and prolongs shelf life in a host of applications. It is also widely used as a laxative and, based on the same induced hyperosmotic effect, in cough syrups (elixirs) and expectorants.

In personal care products glycerol serves as an emollient, humectant, solvent, and lubricant in an enormous variety of products, including toothpaste, where its good solubility and taste give it the edge on sorbitol. Toothpastes are estimated to make up almost one-third of the personal care market for glycerol. Related applications include mouthwashes, skin care products, shaving cream, hair care products and soaps. It is for example a component of "glycerin soap" (Figure 1.6), which is used by people with sensitive, easily irritated skin because its moisturizing properties prevent skin dryness. In general, however, very low concentrations (0.05–1%) are employed, which are not able to significantly reduce the large surplus of biodiesel-generated glycerol on the market.

Glycerol is similar in appearance, smell and taste to diethylene glycol (DEG), which has often been used as a fraudulent replacement for glycerol. In the USA, for example, the Food, Drug, and Cosmetic Act was passed in 1938 following the "elixir sulfanilamide" incident, which caused more than 100 deaths due to the contamination of medicines

Figure 1.6   Glycerin soap.
(Photo courtesy of A K A Saunders, Inc.)

Figure 1.7 Early FDA laboratory.
(Photo courtesy of FDA History Office.)

with DEG. The revamped 1938 law provided food standards, addressed the safety of cosmetics, and required that drugs were checked for safety before sale, reinforcing the role of the Food and Drugs Administration (Figure 1.7).[15] As late as 2007 the Food and Drug Administration blocked all shipments of toothpaste from China to the USA after reports of contaminated toothpaste entering *via* Panama. The toothpaste contained DEG, killing at least 100 people. The poison, falsely labeled as glycerol, had in 2006 been mistakenly mixed into medicines in Panama, resulting in the occurrence of the fatal poisonings. The DEG had originated from a Chinese factory which had deliberately falsified records in order to export DEG in place of the more expensive glycerol.[16] Eventually, a large batch of toothpaste contaminated with DEG also reached the EU market, with a number of poisoning cases being reported in Italy and southern Europe.

The glycerol market is currently undergoing radical changes, driven by very large supplies of glycerol arising from biodiesel production. Researchers and industry have been looking at new uses for glycerin to replace petrochemicals as a source of chemical raw materials, and in a relatively few years there have been an impressive series of achievements. These topics are discussed in the following chapters. After the sustained period of increasing oil prices starting in the early 2000s, glycerol is now becoming established as a major platform for the production of chemicals and fuels.

1.3 Production of Bioglycerol

Glycerol provides the molecular skeleton of all animal and vegetable fats (triglycerides, the energy reservoir for materials in nature). It is also the oldest organic molecule isolated by man, obtained by heating fats in the presence of ash to produce soap as early as 2800 BC.[17] It constitutes on average about 10% by weight of fatty matter. When the body uses fat stores as a source of energy, glycerol and fatty acids are released into the bloodstream. The glycerol component is converted to glucose in the liver and provides energy for cellular metabolism. Natural glycerol is obtained hydrolytically from fats and oils during soap and fatty acid manufacture, and by transesterification (an interchange of fatty acid groups with another alcohol) during the production of biodiesel fuel (Figure 1.8). It therefore comes as no surprise that in these energy intensive days glycerol has become a hot topic in industry at large.

In a certain sense, glycerol was already a national defense priority in the days leading up to World War II, as the supply of glycerol originating from soap making was largely insufficient to meet the wartime demand for nitroglycerine, *i.e.,* for dynamite, the smokeless gunpowder

Figure 1.8 Cover of the September 2006 issue of *Biodiesel Magazine.* (Photo courtesy of the publisher.)

used in all types of munitions, discovered by Swedish industrialist Alfred Nobel. Nobel built bridges and buildings in Stockholm and while researching new methods for blasting rock, he invented in 1863 the detonator for igniting nitroglycerine by means of a strong shock rather than by heat combustion. Nitroglycerine, invented by Italian chemist Ascanio Sobrero in 1846, is very volatile in its natural liquid state. In 1866 Nobel discovered that mixing nitroglycerine with silica (kieselguhr) would convert the liquid into a malleable paste, called dynamite, which could be kneaded and shaped into rods suitable for insertion into drilling holes.

Clearly the supply of glycerol became a strategic war priority for industrial nations worldwide. For example, when World War I began in 1914 DuPont was the only company in the USA which manufactured smokeless powder and was the nation's leading producer of dynamite.[18] Soon after World War I the US Government therefore decided to make its production independent of soap manufacture in order to meet wartime demand, and instead based it on high yield reactions using petroleum feedstock. Since that time glycerol has been produced from epichlorohydrin obtained from propylene, and thus from fossil oil. Today, however, glycerol plants of this type are closing and being replaced by other plants which use glycerol as a raw material, even for the production of epichlorohydrin itself (see Chapter 4).[19] This is the result of the large surplus of glycerol created as a by-product in the manufacture of biodiesel fuel by transesterification of seed oils with methanol using KOH as a base catalyst, or batch esterification of fatty acids catalyzed by sulfuric acid (Figure 1.9).[20]

In the traditional manufacturing process biodiesel is produced by a transesterification reaction between vegetable oil and methanol, catalyzed by KOH. It is an equilibrium reaction with the following stoichiometry (Figure 1.10):

$$100\,kg \text{ of oil} + 10.5\,kg\,MeOH = 100\,kg \text{ methyl esters (biodiesel)}$$
$$+ 10.5\,kg \text{ glycerol} \qquad (1)$$

Colza, soybean and palmitic oils have the most suitable physico-chemical characteristics for transformation into biodiesel. Raw vegetable oil is first refined by degumming (elimination of lecithins and phosphorus) and deacidification (elimination of free fatty acids). The fatty acids comprise some 2% of the original product, and after distillation are recovered and sold as by-products. The oil is charged into large batch reactors and heated at 55 °C with a 30% excess of a mixture of methanol and KOH. After reaction for 2 h the mixture is left to stand.

Figure 1.9   A batch of home-made canola oil biodiesel sitting on its glycerol layer. (Photo courtesy of Biodieselcommunity.org.)

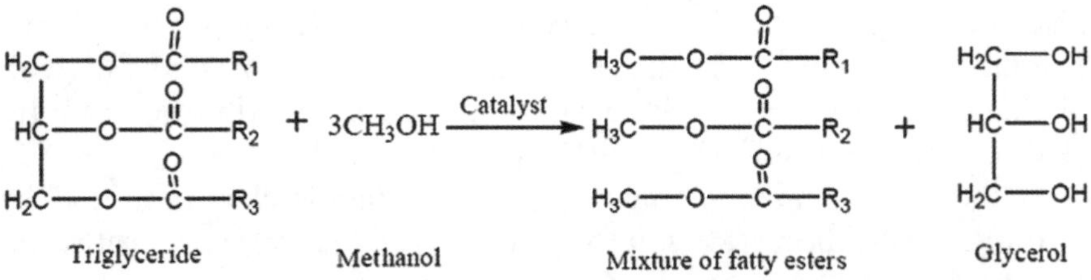

Figure 1.10   Acylglycerol transesterification with methanol.

The glycerol–methanol solution is heavier than methanol and the esters, and is run off from the bottom of the reactor (Figure 1.9). Methanol is expensive and is recovered and reused in a further reaction run. The remaining mixture, comprizing biodiesel, glycerol–methanol solution and methanol, is distilled to complete the recovery of the methanol and washed and centrifuged to eliminate traces of glycerol. The resulting product, a mixture of biodiesel and water, is dried under vacuum and put into stock pending analytical tests. Fundamental analysis parameters are the ester content (minimum 96.5%) and free glycerol content (maximum 200 ppm).

The glycerol side stream typically contains a mixture of glycerol, methanol, water, inorganic salts (catalyst residue), free fatty acids, unreacted mono-, di-, and triglycerides, methyl esters, and a variety of other "matter organic non-glycerol" (MONG) in varying proportions. The methanol is typically stripped from this stream and reused, leaving crude glycerol after neutralization. In its raw state crude glycerol has a high salt and free fatty acid content and a substantial color (yellow to dark brown). Consequently, crude glycerol has few direct uses, and its fuel value is also marginal. An economic solution for the purification of crude glycerol streams combines electrodialysis and nanofiltration, affording a colorless liquid with low salt content, equivalent to technical grade purity (Figure 1.11).[21] The recovered glycerol, after polishing if necessary using ion exchange and removal of water–methanol solution by evaporation, easily meets USP glycerol standards. This membrane-based technique avoids expensive evaporation and distillation, and also such problems as foaming, carry-over of contaminants, and limited recovery.

In general, the traditional biodiesel manufacturing processes have several disadvantages, including soap formation (from the KOH catalyst); use of an excess of alcohol (for shifting the equilibrium to fatty esters), which must be separated and recycled; homogeneous catalysts which require neutralization, causing salt waste streams; the expensive separation of products from the reaction mixture; and relatively high investment and operating costs. In 2005, the Institut Français du Pétrole (IFP) disclosed a novel biodiesel process, called Esterfif. Starting from

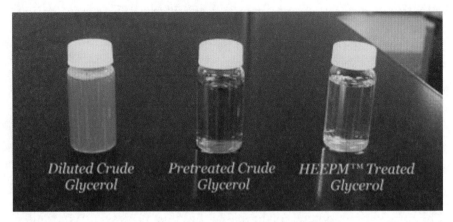

Figure 1.11 Glycerol purified by membranes combining electrodialysis and nanofiltration.
(Photo courtesy of EET Corporation.)

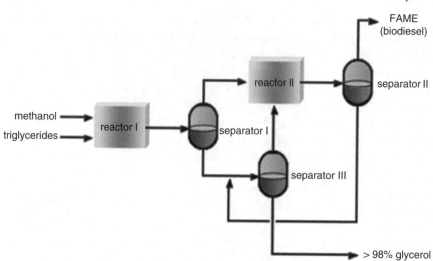

Figure 1.12 Simplified schematic of the IFP Esterfif biodiesel process, based on two consecutive reactor–separator stages. (Reproduced from Ref. 23, with permission.)

triglycerides, the transesterification step was performed using a solid catalyst, a mixed Zn–Al oxide.[22] The process runs at higher temperature and pressure than the homogeneous method and uses an excess of methanol, which is vaporized and recycled. It has two reactors and two separators, needed for shifting the methanolysis equilibrium (Figure 1.12). At each stage, the excess of methanol is removed by partial evaporation and the esters and glycerol are separated in a settler.

A second alternative, developed by the Dutch company Yellowdiesel in 2006,[24] is especially suited to mixed feedstocks with high free fatty acid (FFA) content, such as those used for cooking oil and low-grade greases. The process combines the reaction and the separation in a single step, using reactive distillation (also known as catalytic distillation, Figure 1.13).[25] This intensifies mass transfer, allows *in situ* energy integration, reduces equipment costs, and simplifies process flow and operation.[26]

Furthermore, the thermodynamic equilibrium of the reaction can be shifted by controlling the vapour–liquid equilibrium in the column. The first pilot biodiesel plant based on this process (2500 tonnes/year), built by Fertibom, will come on stream in Rio de Janeiro in 2008.

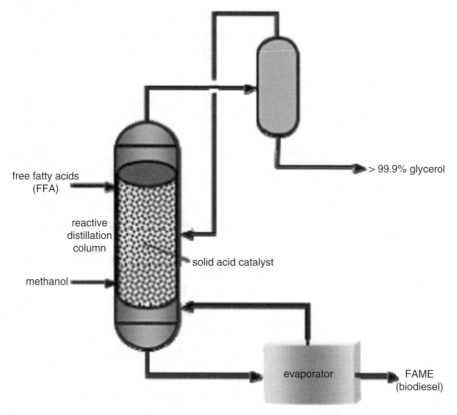

Figure 1.13 Simplified schematic of the Yellowdiesel catalytic distillation process for making biodiesel from high-FFA oils, by integrating the reaction and separation within one reactive distillation column using a solid acid catalyst. (Reproduced from Ref. 23, with permission.)

References

1. J. Bonnardeaux, *Glycerin Overview*, Report for the Western Australia Department of Agriculture and Food. November 2006. http://www.agric.wa.gov.au/content/sust/biofuel/glycerinoverview.pdf.
2. *CRC Handbook of Chemistry and Physics*, 87th edn., Boca Raton (FL): 2006.
3. M. A. David, G. S. Henry Academy, *Glycerol: A Jack of all Trades*. This nice 1996 essay can be accessed at the URL: http://www.chem.yorku.ca/hall_of_fame/essays96/glycerol.htm#litharge.
4. C. S. Callam, S. J. Singer, T. L. Lowary and C. M. Hadad, Computational analysis of the potential energy surfaces of glycerol in the gas and aqueous phases: Effects of level of theory, basis set, and solvation on strongly intramolecularly hydrogen-bonded systems. *J. Am. Chem. Soc.*, 2001, **123**, 11743.

5. R. Chelli, P. Procacci, G. Cardini and S. Califano, Glycerol condensed phases. Part II: A molecular dynamics study of the conformational structure and hydrogen bonding. *Phys. Chem. Chem. Phys.*, 1999, **1**, 879.
6. R. Zondervan, F. Kulzer, G. C. G. Berkhout and M. Orrit, Local viscosity of supercooled glycerol near T_g probed by rotational diffusion of ensembles and single dye molecules. *Proc. Natl. Acad. Sci.* 2007, **104**, 12628.
7. S. L. Bradbury and W. B. Jakoby, Glycerol as an enzyme-stabilizing agent: Effects on aldehyde dehydrogenase. *Proc. Natl. Acad. Sci.*, 1972, **69**, 2373.
8. R. Grandori, I. Matecko, P. Mayr and N. Müller, Probing protein stabilization by glycerol using electrospray mass spectrometry. *J. Mass Spectrom.*, 2001, **36**, 918.
9. I. Gill and A. Ballesteros, Encapsulation of biologicals within silicate, siloxane, and hybrid sol–gel polymers: An efficient and generic approach. *J. Am. Chem. Soc.*, 1998, **120**, 8587.
10. N. Nassif, O. Bouvet, M. Noelle Rager, C. Roux, T. Coradin and J. Livage, Living bacteria in silica gels. *Nature Mater.*, 2002, **1**, 42.
11. D. Bohm: *Wholeness and the Implicate Order*, Oxford: Routledge, 1996. Underlying the apparently chaotic realm of physical appearances – the explicate order – there is always a deeper, implicate order that is often hidden.
12. We thank Prof. A. Helman, University of California, Santa Cruz, for allowing reproduction of this figure from the Physics Lecture Demonstration Catalog: http://physics.ucsc.edu/lecture demonstrations.
13. See the recent report from Frost & Sullivan: "R&D creating new avenues for glycerol" (4 August 2006). Available online at the URL: https://www.frost.com/prod/servlet/market-insight-top.pag?docid=77264824.
14. The term kosher refers to food recognized as such by the Torah. The laws of kosher detail the permitted and forbidden animals, fish, and fowl, and describe the separation of dairy and meat. Jewish rabbis certify food products accordingly, including glycerol for cough syrups. To be Kashrut endorsed, glycerol must be from vegetable sources instead of from animal tallow.
15. F. Case, 100 years of the FDA. *Chemistry World* (July 2006). Similar pictures are available at the URL: http://www.rsc.org/chemistryworld/restricted/2006/july/100yearsfda.asp.
16. S. Reinberg, FDA bans toothpaste from China. *Washington Post*, June 1, 2007.

17. J. A. Hunt, A short history of soap. *Pharm. J.*, 1999, **263**, 985.

18. In two years DuPont sales increased from $25 million to $318 million and profits soared from $5.6 million to $82 million. DuPont used these profits to diversify into dyestuffs, plastics and paints.

19. (a) The US agribusiness Archer Daniels Midland recently announced plans to make propylene glycol from glycerol, instead of from propylene oxide, in a process employing advanced catalysts. Dow Chemical closed its glycerol plant in Texas early this year when Procter & Gamble Chemicals shut down a natural glycerol refinery in England. See M. McCoy, Glycerin Surplus. *Chem. Eng. News* 2006, **84** (6), 7.

20. (a) M. A. Harmer, W. E. Farneth and Q. Sun, Towards the sulfuric acid of solids. *Adv. Mater.*, 1998, **10**, 1255; (b) E. Lotero, Y. Liu, D. E. Lopez, K. Suwannakarn, D. A. Bruce and J. G. Goodwin, Synthesis of biodiesel via acid catalysis. *Ind. Eng. Chem. Res.*, 2005, **44**, 5353.

21. This efficient electro-pressure membrane (HEEPM) technology for desalinating liquids is available from the EET Corporation: http://www.eetcorp.com/heepm/glycerine.htm.

22. L. Bournay, D. Casanave, B. Delfort, G. Hillion and J. A. Chodorge, New heterogeneous process for biodiesel production: A way to improve the quality and the value of the crude glycerin produced by biodiesel plants. *Catal. Today*, 2005, **106**, 190.

23. G. Rothenberg, *Catalysis: Concepts and Green Applications*, Weinheim: Wiley–VCH, 2008, ISBN 978-3-527-31824-7.

24. (a) A. A. Kiss, A. C. Dimian and G. Rothenberg, Solid acid catalysts for biodiesel production – towards sustainable energy. *Adv. Synth. Catal.*, 2006, **348**, 75; (b) A. A. Kiss, F. Omota, A. C. Dimian and G. Rothenberg, The heterogeneous advantage: Biodiesel by catalytic reactive distillation. *Top. Catal.*, 2006, **40**, 141.

25. (a) F. Omota, A. C. Dimian and A. Bliek, Fatty acid esterification by reactive distillation: Part 2 – kinetics-based design for sulphated zirconia catalysts. *Chem. Eng. Sci.*, 2003, **58**, 3175; (b) F. Omota, A. C. Dimian and A. Bliek, Fatty acid esterification by reactive distillation. Part 1: Equilibrium-based design. *Chem. Eng. Sci.*, 2003, **58**, 3159.

26. (a) H. Subawalla and J. R. Fair, Design guidelines for solid-catalyzed reactive distillation systems. *Ind. Eng. Chem. Res.*, 1999, **38**, 3696; (b) H. G. Schoenmakers and B. Bessling, Reactive and catalytic distillation from an industrial perspective. *Chem. Eng. Prog.*, 2003, **42**, 145.

CHAPTER 2
Aqueous Phase Reforming

2.1 Glycerol as a Platform for Green Fuels

From both industrial and innovation viewpoints, the major achievement of the new glycerol chemistry is the aqueous phase reforming process (APR), in which glycerol in the aqueous phase is converted to hydrogen and carbon monoxide (the synthesis gas, or "syngas") under relatively mild conditions (temperatures between 225 and 300 °C) using a Pt–Re catalyst in a single reactor.[1] This formation of synthesis gas is crucial for the bio-refinery, since the gas can be used as a source of fuels and chemicals using the Fischer–Tropsch (or methanol) synthesis, offering an energy-efficient alternative to liquid transportation fuels from petroleum (Figure 2.1).

Similarly, an equally mild APR process using a Pt catalyst affords the rapid production of high yields of hydrogen fuel from glycerol at very low CO concentrations, due to more favorable water–gas shift (WGS) thermodyamics, and with considerably lower energy consumption than traditional methane reforming (Figure 2.2).[2]

Hydrogen from renewable sources with low CO content is urgently needed for fuelling the highly efficient fuel cells that will soon find widespread use for the heating of buildings, hospitals and factories, and probably also for cars and other vehicles. Petroleum currently provides a significant fraction (37%) of the world's energy.[4] Almost 70% of this is consumed by the transportation sector, which at present relies entirely on petroleum for its energy needs.[5] Biomass is the main candidate as an alternative source of transportation fuel, since it is renewable and CO_2 neutral (Figure 2.3).[6] The quantity of biomass grown annually is suffi-cient to provide energy for approximately 70% of the transportation sector, provided it can be converted to clean-burning fuels having high energy density, as is currently the case with petroleum.[7] Biomass is comprised primarily of lignin and carbohydrates (*e.g.*, starch and

RSC Green Chemistry Book Series
The Future of Glycerol: New Uses of a Versatile Raw Material
By Mario Pagliaro and Michele Rossi
© Mario Pagliaro and Michele Rossi 2008

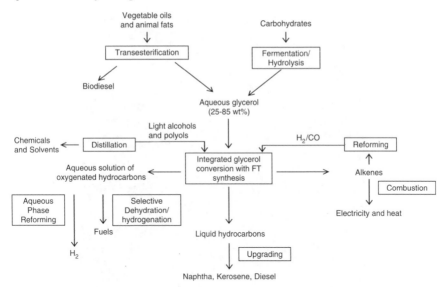

Figure 2.1 Process pathway for production of liquid fuels from biomass by integrated glycerol conversion to synthesis gas and Fischer–Tropsch synthesis. (Reproduced from Ref. 1, with permission.)

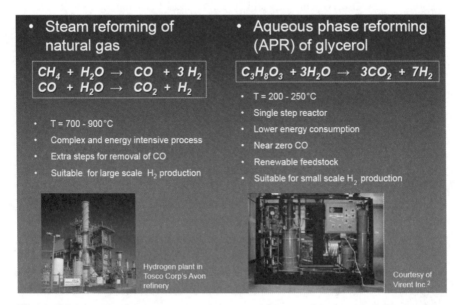

Figure 2.2 Production of hydrogen by aqueous phase reforming of glycerol is more convenient than steam reforming of natural gas. (Reproduced from Ref. 2, with permission.)

Figure 2.3   The first commercial scale Fischer–Tropsch plant using biomass, with a capacity of just over 4000 barrels per day, is planned to begin operation in Germany after 2008. The Fischer–Tropsch process (here shown in a plant at Sasol) can produce a high quality diesel fuel from biomass. The fuel is competitive if oil costs are above $40 a barrel. (Reproduced with the permission of Sasol Corporation, South Africa.)

cellulose),[3] and one method of converting these to liquid fuels is by fermentation to liquid alcohols such as ethanol and butanol. However, the overall energy balance for production of bioethanol from grain-derived starches by a combination of hydrolysis, fermentation, and distillation[5] is not favourable, in fact it has been estimated that the amount of energy required to produce bioethanol is slightly greater than the energy content of the ethanol produced (energy return on investment (EROI) = 1.15).[8]

On the other hand, the energy balance for the coupled APR and FT reactions (Figure 2.4) is favorable. Formation of synthesis gas from glycerol is highly endothermic with an enthalpy change of about 80 kcal/mol, but the conversion of synthesis gas to alkanes (along with CO_2 and water) is highly exothermic (−110 kcal/mol). This means that the conversion of glycerol to alkanes by a combination of reforming and Fischer–Tropsch synthesis is mildly exothermic overall, with an enthalpy change and an overall gain in energy of about −30 kcal/mol of glycerol. This process therefore provides the opportunity for improving the economic viability of biomass-based Fischer–Tropsch synthesis

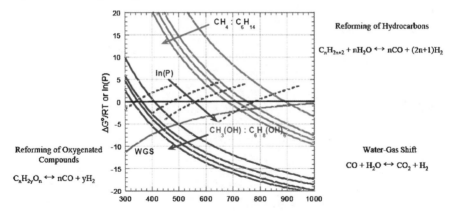

Figure 2.4 APR of polyols is thermodynamically favorable at significantly lower temperature than steam reforming of methane. (Reproduced from Ref. 2, with permission.)

by reducing the cost of synthesis gas production and by improving its thermal efficiency. Glycerol can in fact be obtained by the fermentation of glucose,[9] affording a 25% solution of glycerol; this offers an energy-efficient alternative to ethanol-based production, since higher product concentrations can be obtained. On the other hand the fermentation process used to produce bioethanol from carbohydrates leads to an aqueous solution containing only 5–10 wt% ethanol, and significant expenditure of energy is required to produce fuel grade ethanol by distillation of such a dilute solution.[10]

Such improvements in synthesis gas generation and thermal efficiency to improve the economics of the Fischer–Tropsch synthesis are crucial. The low thermal efficiency and high operating costs of this process, of which more than 50% stems from synthesis gas generation,[11] has traditionally made liquid alkane production by this synthesis economically viable only on the large scale. And this holds true equally for both conventional Fischer–Tropsch synthesis (*i.e.*, using coal or natural gas), and also for the "green" Fischer–Tropsch synthesis involving advance gasification of biomass to form the synthesis gas.[12]

2.2 Production of Hydrogen *via* Aqueous Phase Reforming

The APR process efficiently generates easily purified hydrogen by the reaction of glycerol with water to form carbon dioxide and hydrogen (Equation 1):

$$C_3H_8O_3 + 3H_2O \rightarrow 3CO_2 + 7H_2 \tag{1}$$

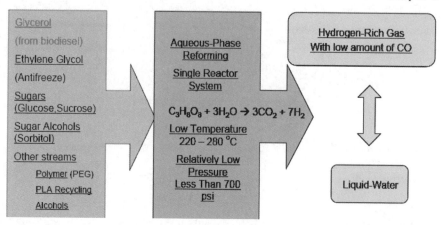

Figure 2.5 The APR process efficiently generates hydrogen from polyols. For bio-mass-generated substrates, the carbon released as carbon dioxide is that derived from plants.

This reforming reaction takes place over a catalyst in a single reactor at temperatures between 200 and 250 °C and at pressures above the bubble point of water (16 and 40 bar, respectively). The gas stream leaving the APR can be utilized directly as a high energy fuel for internal combustion engines, gas-fired turbines or solid oxide fuel cells (Figure 2.5). The worldwide market for hydrogen is estimated to be approximately 45 million tonnes annually. In addition to its use as an energy provider in transportation applications, hydrogen is a key chemical building block in many chemical processes. This is particularly the case in ammonia fertilizer production, and also in oil refineries for upgrading lower quality oil fractions into gasoline and diesel and to remove sulphur contaminants. It is additionally important in the manufacture of glass, vitamins, personal care products, lubricants, refined metals and food products.

The process for the production of hydrogen from glycerol is cost effective since:

- it generates hydrogen without the need to volatilize water, which represents a major energy saving;
- it occurs at temperatures and pressures where the water–gas shift reaction is favorable, making it possible to generate hydrogen with low quantities of CO in a single chemical reactor;
- it occurs at pressures, typically 15 to 50 bar, at which the hydrogen-rich effluent can be purified effectively using pressure-swing adsorption (PSA) technology; and

- it takes place at low temperatures, which minimizes the undesirable decomposition reactions typically encountered when carbohydrates are heated to elevated temperatures.

The choice as a starting point for hydrogen production decreases in the order: glycerol > sorbitol > glucose. The raw water-soluble glycerol waste from biodiesel manufacture is an ideal feedstock for the APR process. At the reaction temperatures investigated CO concentrations were below 300 ppm, and this method is particularly well suited for generating hydrogen on demand in a compact and highly efficient single-step reactor (Figure 2.6). Operating temperatures are one-third of those in other processes, with the generation of 10 times more hydrogen per g of catalyst. In the original study,[2] the reaction of glycerol was conducted over a precious metal catalyst supported on γ-alumina. Subsequent investigations have shown that a range of catalyst compositions are active for the generation of hydrogen by aqueous phase reforming of oxygenated compounds, in particular inexpensive nickel-based materials and selective APR catalysts.[13]

The noncondensable gas stream leaving the APR contains predominately carbon dioxide and hydrogen. Hydrogen is readily purified from this gas stream utilizing PSA technology. Importantly, the gas stream leaving the APR is within the required feed pressure range for the PSA unit (between 16 and 40 bar). Accordingly, the PSA unit does not need an expensive and energy-consuming compressor to provide the necessary feed pressure. This results in lower capital costs and increased energy efficiency.

The PSA technology generates a waste hydrogen stream (typically 10–20% of the feed) due to the pressure swing and purging cycles, which

Figure 2.6 An aqueous phase reforming reactor is small and locally generates hydrogen on demand.
(Photo courtesy of Virent Inc.)

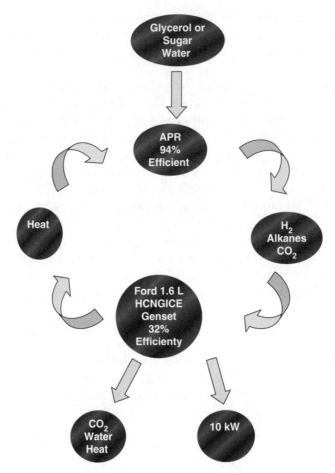

Figure 2.7   The APR reactor at the MGE plant uses heat generated by the combustion
of hydrogen and methane to generate electricity. (Adapted from Ref. 16,
with permission.)

also contain the alkanes generated in the APR process. Combustion of
the waste hydrogen and alkanes can provide much of the process heat
necessary for the reactor, because the reforming reaction is endothermic
and requires heat input. Since the APR process runs at low temperature,
an alternative method of providing this process energy is to recycle the
high temperature waste heat, as shown in Figure 2.7.

2.3   Production of Hydrocarbon Fuels *via* Aqueous Phase Reforming

Liquid alkanes can be produced directly from glycerol in a two-bed
reactor system using an integrated process consisting of APR followed

in a single reactor by Fisher–Tropsch conversion.[14] Glycerol can be converted in this way to synthesis gas at high rates and selectivity at temperatures between 225 and 335 °C, according to Equation 2.

$$C_3O_3H_8 \rightarrow 3CO + 4H_2 \tag{2}$$

Operation at low temperatures provides the opportunity to couple this endothermic glycerol conversion with the exothermic Fischer–Tropsch synthesis to produce liquid transportation fuel from aqueous glycerol solutions *via* an integrated process (Equation 3):[14]

$$C_3O_3H_8 \rightarrow \frac{7}{25}C_8H_{18} + \frac{19}{25}CO_2 + \frac{37}{25}H_2O \tag{3}$$

In particular, either glycerol conversion or Fischer–Tropsch synthesis can be carried out effectively under the same conditions in a two-bed reactor system (Figure 2.8).

This integrated glycerol-based process improves the economics of "green" Fischer–Tropsch synthesis by reducing costs, by eliminating the

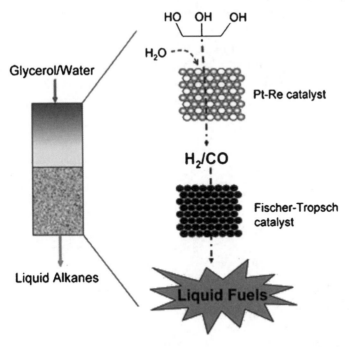

Figure 2.8 The integrated process for making fuels from glycerol is a simple two-step catalytic process carried out at low temperature and moderate pressure, and effectively harnesses the energy from a renewable resource. (Reproduced from the RSC website, with permission.)

need for an a biomass gasifier, by reducing the size of the synthesis reactor, by producing an undiluted synthesis gas stream and by eliminating subsequent cleaning steps. In addition, the process can produce synthesis gas of varying H_2–CO composition, thus eliminating the need for a water–gas shift reactor and allowing for the use of Fischer–Trospch catalysts that operate at different H_2:CO ratios. Synthesis gas production and clean-up are indeed the critical steps in producing liquid alkanes from biomass and account for more than 50% of the total costs of producing these by "green" Fischer–Tropsch synthesis.

Hence, for example, conversion of glycerol over a 10 wt% Pt–Re/C catalyst with a Pt:Re ratio of 1:1 produces a synthesis gas stream suitable for Fischer–Tropsch synthesis (H_2:CO between 1.0 and 1.6) from concentrated glycerol feed solutions at low temperatures (275 °C) and pressures up to 17 bar, where incomplete vaporization of the glycerol feed occurs (Figure 2.9). The synthesis gas produced in the first process is fed directly into the second reactor consisting of a 2.9 wt% Ru–TiO_2 catalyst bed whose intermediate value of χ ($50 \times 10^{16}\,m^{-1}$) leads to optimum C_{5+} selectivity.

This combined process produces liquid alkanes, with S_{C5+} between 0.63 and 0.75 at 275 °C and pressures between 5 and 17 bar, with more than 40% of the carbon in the products contained in the organic liquid phase at 17 bar. The aqueous liquid effluent from the integrated process contains between 5 and 15 wt% methanol, ethanol and acetone, which can be separated from the water by distillation and reused, or recycled

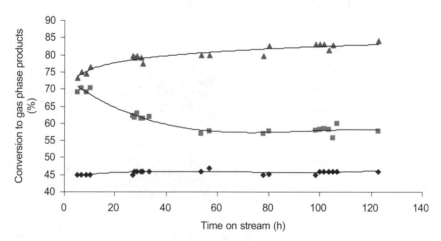

Figure 2.9   Conversion to gas phase products (♦), CO/CO_2 molar ratio (●), and H_2/CO molar ratio (■) for gas phase processing of 30 wt% aqueous glycerol feedstock at 548 K and 8.3 bar. (Adapted from Ref. 14, with permission.)

for conversion to gaseous products. Importantly, the coupling of glycerol conversion to synthesis gas and Fischer–Tropsch synthesis leads to synergies in the operations of these processes, such as

- avoiding the highly endothermic and exothermic steps that would result from the separate operation of these processes;
- eliminating the need to condense water and oxygenated hydrocarbon by-products between the catalyst beds;
- allowing operation at higher pressures (*i.e.*, 17 bar), at which synthesis gas production over Pt–Re/C is decreased and the production of liquid by-products increased; and
- causing an increase in selectivity to C_{5+} hydrocarbons.

The primary oxygenated hydrocarbon intermediates formed during conversion of glycerol to synthesis gas are ethanol, acetone, and acetol; all have positive effects on the Fischer–Tropsch synthesis step. In particular, water, ethanol and acetone have a slightly positive effect, such as increasing the selectivity to C_{5+} hydrocarbons (S_{C5+}) by a factor of 2 (from 0.30 to 0.60). Acetol can participate in Fischer–Tropsch chain growth, forming pentanones, hexanones and heptanones in the liquid organic effluent stream.

Figure 2.10 shows the product molecular weight distributions for experiments that combined glycerol conversion with Fischer–Tropsch

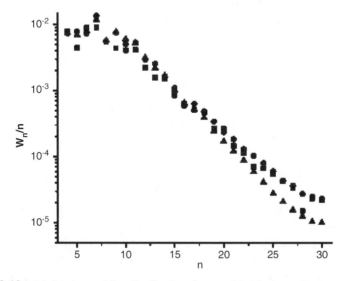

Figure 2.10 Molecular weight distribution for combined glycerol conversion with Fischer–Tropsch synthesis experiments at 548 K, and 5 bar (■), 11 bar (●), and 17 bar (▲). (Reproduced from Ref. 14, with permission.)

synthesis, and these distributions exhibit deviations from the kinetics of the Anderson–Schulz–Flory (ASF) chain growth model similar to the Fischer–Tropsch experiments, indicating α-olefin readsorption effects. The oxygenated hydrocarbon by-products from glycerol react over the Ru–TiO$_2$ bed, most likely by entering into chain growth. The C$_{5+}$ selectivity, selectivity to pentanones, hexanones and heptanones in the organic liquid, and the conversion of CO for combined glycerol conversion with Fischer–Tropsch synthesis at 11 and 17 bar, are all higher than those at 5 bar, despite the fact that synthesis gas production from glycerol is decreased at these elevated pressures. The more favorable Fischer–Tropsch conditions (*i.e.*, higher pressure) are more important to the integrated process than the synthesis gas production rate.

2.4   Industrial Applications

In the Netherlands, BioMethanol Chemie Nederland uses glycerol to produce synthesis gas, which is then reformed to make methanol. The consortium has reopened the Methanor production facilities in Delfzijl after the plant was taken over from the previous owners. Production of biomethanol will start at the end of 2007 at 20 000 tonnes/year, and is expected to grow to 1 million tonnes/year by 2010.[15] The APR technology has a large commercial potential, which has been reinforced by coupling with the Fisher–Tropsch fuel production. The different product streams of the integrated APR–FT process each have potential end uses (Figure 2.1). The most likely use for the gas alkanes would be combustion to produce process heat and electricity, with some of the unconverted hydrogen and carbon monoxide being recycled to the Fischer–Tropsch bed.

The aqueous liquid product stream containing oxygenated hydrocarbons (*e.g.*, ethanol, methanol and acetone) at concentrations between 5–15 wt% is suitable for distillation (oxygenated hydrocarbons are intermediates and solvents in the chemical industry). Indeed, the process has spawned the company Virent Energy Systems Inc, created by its inventors in 2002 to commercialize the APR process. Having labelled "BioForming" the process which enables the production of renewable fuels and chemicals from glycerol and carbohydrates, in 2005 Virent contracted with Madison Gas and Electricity (MGE), a local utility, to build an integrated BioForming reactor that converts glycerol into hydrogen and methane and burns the mixture to drive a 10 kW generator for electricity production. The system (Figure 2.7) has delivered power to the MGE grid since its start-up at the beginning of 2006. It also utilizes the waste heat from the internal combustion engine to provide the limited

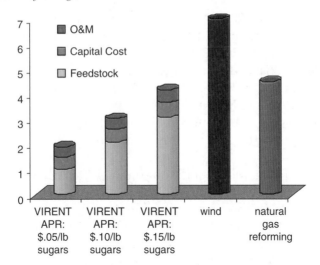

Figure 2.11 As cost of the feedstock is the primary driver of the overall cost of hydrogen generated from biomass-derived hydrocarbons, increasing biodiesel production will make its production from glycerol a commercial reality. (Adapted from Ref. 16, with permission.)

amount of process heat required for the reactor. As a result, the thermal efficiency for this initial unit is excellent. The APR reactor initially used pure glycerol, but it was also designed to run using glycerol in crude form.

The success of this system led large companies such as Cargill, Shell and Honda to invest in the company, as they were interested in seeing how the technology could be used to supply renewable hydrogen for fuel cells. Collaboration with Shell Hydrogen is expected to speed the time of the technology to market. The production of gasoline *via* APR in June 2006 confirmed that the technology was a viable pathway to the production of liquid fuels and chemicals currently derived from fossil fuels. Biodiesel-generated glycerol is a cost-effective feedstock and positions the APR process to compete effectively with conventional fuels, even before consideration of tax incentives for renewable. As can be seen in Figure 2.11, the primary economic driver for the APR process is the feedstock.[16] A comparable steam reformer utilizing non-renewable natural gas is expected to be 56% efficient, and could generate hydrogen at a cost of $4.50 per kg in a distributed system.

References

1. R. R. Soares, D. A. Simonetti and J. A. Dumesic, Glycerol as a source for fuels and chemicals by low-temperature catalytic processing. *Angew. Chem. Int. Ed.*, 2006, **45**, 3982.

2. R. D. Cortright, R. R. Davda and J. A. Dumesic, Hydrogen from catalytic reforming of biomass-derived hydrocarbons in liquid water. *Nature*, 2002, **418**, 964.

3. B. Liu, Y. Zhang, J. W. Tierney and I. Wender, Hydrogen generation from glycerol via aqueous phase reforming, Presentation at the CFFS Annual Meeting, Daniels (WV), August 1, 2006.

4. G. Alexander and G. Boyle, Introducing renewable energy, in *Renewable Energy*, ed. G. Boyle, Oxford University Press, New York, 2004.

5. Energy Information Administration, US Department of Energy, http://www.eia.doe.gov.

6. D. L. Klass, *Biomass for Renewable Energy, Fuels, and Chemicals*, Academic Press, San Diego, 1998.

7. G. W. Huber, S. Iborra and A. Corma, Synthesis of transportation fuels from biomass: Chemistry, catalysts, and engineering. *Chem. Rev.*, 2006, **106**, 4044.

8. J. Hill, E. Nelson, D. Tilman, S. Polasky and D. Tiffan, Environmental, economic, and energetic costs and benefits of biodiesel and ethanol biofuels. *Proc. Natl. Acad. Sci.*, 2006, **103**, 11206.

9. C. S. Gong, J. X. Du, N. J. Cao and G. T. Tsao, Co-production of ethanol and glycerol. *Appl. Biochem. Biotechnol.*, 2000, **84–86**, 543.

10. A. E. Farrell, R. J. Plevin, B. T. Turner, A. D. Jones, M. O'Hare and D. M. Kammen, Ethanol can contribute to energy and environmental goals. *Science*, 2006, **311**, 506.

11. P. L. Spath and D. C. Dayton, *Preliminary screening—Technical and economic assessment of synthesis gas to fuels and chemicals with emphasis on the potential for biomass-derived syngas*, US Department of Energy, National Renewable Energy Laboratory, NREL/ TP-510-34929, 2003, 1–142.

12. (a) C. N. Hamelinck, A. P. C. Faaij, H. den Uil and H. Boerrigter, Production of FT transportation fuels from biomass; technical options, process analysis and optimisation, and development potential. *Energy*, 2004, **29**, 1743–1771; (b) M. J. A. Tijmensen, A. P. C. Faaij, C. N. Hamelinck and M. R. M. van Hardeveld, Exploration of the possibilities for production of Fischer–Tropsch liquids and power *via* biomass gasification. *Biomass Bioenerg.*, 2002, **23**, 129.

13. G. W. Huber, J. W. Shabaker and J. A. Dumesic, Raney Ni–Sn catalyst for H_2 Production from biomass-derived hydrocarbons. *Science*, 2003, **300**, 2075.

14. D. A. Simonetti, J. Rass-Hansen, E. L. Kunkes, R. R. Soares and J. A. Dumesic, Coupling of glycerol processing with Fischer–Tropsch synthesis for production of liquid fuels. *Green Chem.*, 2007, **9**, 1073.
15. See also the company's website: http://www.biomcn.com.
16. This analysis is at the URL: http://www.ecw.org/biomass2power/economics.htm.

CHAPTER 3
Selective Reduction

3.1   Reduction of Glycerol

The main product arising from the reduction of glycerol is 1,2-propanediol. 1,2-Propanediol is an important commodity chemical traditionally derived from propylene oxide. Bio-routes enable reduction to 1,3-propanediol, an important monomer which can be polymerized with terephthalic acid to produce polyester fibers known as Sorona (DuPont) or Corterra (Shell). Either diol can thus now be produced by selective dehydroxylation of glycerol through chemical hydrogenolysis or biocatalytic reduction. In the words of one industry practitioner, the imminent commercial production of 1,2-propanediol is turning the glycerol glut into an "advantage for the biodiesel industry".[1]

3.2   Hydrogenolysis to Propylene Glycol

Figure 3.1 summarizes the conversion of glycerol to glycols. In the presence of metallic catalysts and hydrogen, glycerol can be hydrogenated to 1,2-propanediol (propylene glycol), 1,3-propanediol, or ethylene glycol.

The production of propylene glycol by glycerol hydrogenolysis is a process used commercially.[2] The method is based on hydrogenolysis over a copper chromite catalyst ($CuO.Cr_2O_3$) at 200 °C and less than 10 bar, coupled with reactive distillation (Figure 3.2).[3]

The reaction pathway proceeds *via* an acetol (hydroxyacetone) intermediate, using a two-step reaction process under mild reaction conditions (Figure 3.3). In the first step relatively pure acetol is produced from glycerol at 200 °C and 0.65 bar pressure in the presence of a copper chromite catalyst. In the second step, the acetol is further hydrogenated to 1,2-propanediol at 200 °C and 13.8 bar hydrogen pressure using a catalyst similar to that in the first step. This affords propylene glycol in

RSC Green Chemistry Book Series
The Future of Glycerol: New Uses of a Versatile Raw Material
By Mario Pagliaro and Michele Rossi

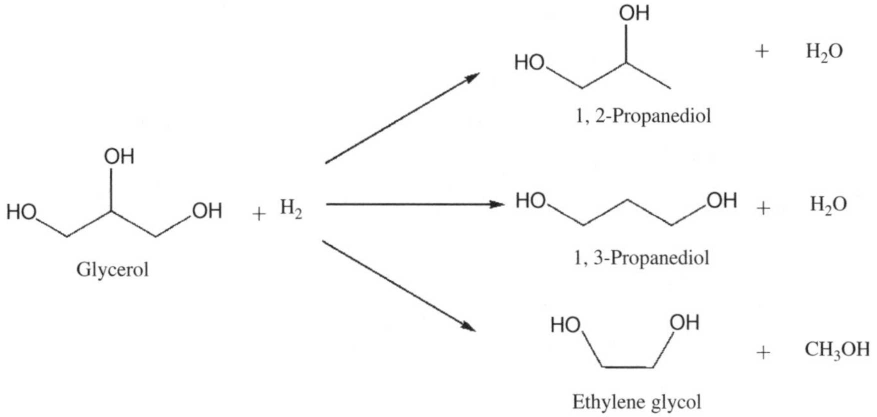

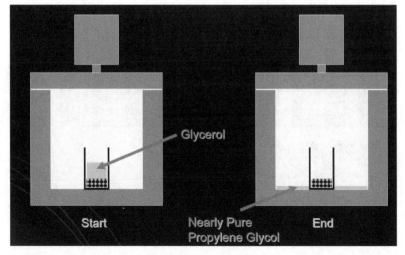

Figure 3.1 The conversion of glycerol to propylene and ethylene glycols.

Figure 3.2 Early observation of the outcomes of glycerol hydrogenolysis over Cu–Cr catalyst showed that the process was actually a reactive distillation. (Reproduced from Ref. 4, with permission.)

>90% yield and at a significantly lower cost than starting from petroleum.[4]

A screening study of catalysts (Figure 3.4, with T = 200 °C and P = 13.8 bar) has shown that good selectivity for propylene glycol with high conversion is readily achieved using a copper chromite catalyst. At temperatures above 200 °C the selectivity to propylene glycol decreases due to excessive hydrogenolysis of the 1,2-propanediol. The process is a true reactive distillation, and researchers have observed the

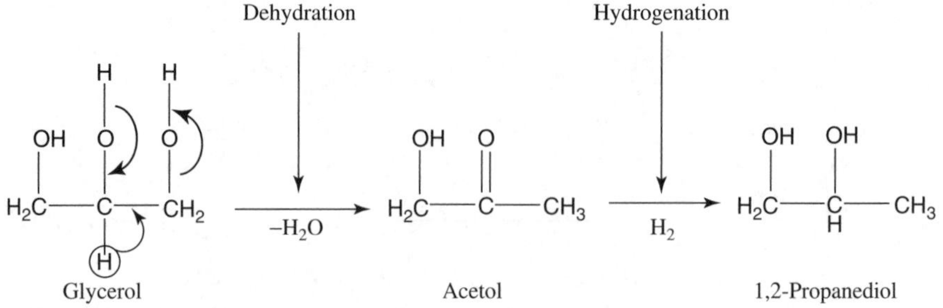

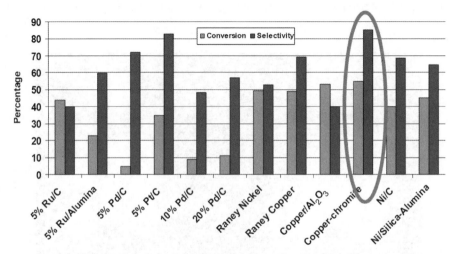

Figure 3.3 Reaction mechanism for conversion of glycerol to propylene glycol. (Reproduced from Ref. 3, with permission.)

Figure 3.4 Commercial catalysts screened for use in glycerol hydrogenolysis. (Reproduced from Ref. 4, with permission.)

delivery of virtually pure propylene glycol from the reaction vessel in the autoclave.

Major problems encountered when the reaction is conducted in a single step are that the catalyst becomes coated with oligomers and that selectivity for propylene glycol above 80% is difficult to achieve. However, if the reactive distillation is conducted in two steps, by combining the reaction and separation steps, the yield of propylene glycol is *ca.* 99% and the life of the catalyst is considerably extended. One feature is that acetol and water are simultaneously removed from the reaction mixture during the heating step as they are formed (glycerol boils at 290 °C and acetol at 140 °C); furthermore, the lower pressure used in the first of the two steps prolongs catalyst life. Further reduction of the acetol–water feed (50% wt) with hydrogen over a similar copper chromite catalyst at 185 °C and 13.8 bar affords 1,2-propylene glycol in >95% selectivity and 99% conversion.

There are two main advantages to this new process. Firstly, the copper chromite catalyst can be used to convert crude glycerol without further purification, whereas supported noble metal catalysts are easily poisoned by contaminants such as chlorides.[5] Secondly, the acetol formed as an intermediate is an important monomer used in industry in the manufacture of polyols. When produced from petroleum acetol costs *ca.* $10 per kg, whereas by hydrogenolysis of glycerol it costs as little as $1 per kg, opening up even more potential applications and markets for glycerol.

Finally, Davy Process Technology (DPT) has developed and patented its own process in which the vapor-phase hydrogenation of glycerol over a heterogeneous copper catalyst yields 1,2-propanediol (Figure 3.5).[6]

Plant development is expected to proceed once 70 000 tonnes per annum if glycerol of reliable quality is available for processing at the right economics. Impressive selectivity for this product (>96%) has been reported using this catalyst system.

There are several routes to propylene glycol from renewable feedstocks. The most common of these is by hydrogenolysis of sugars or sugar alcohols at high temperature and pressure in the presence of a metal catalyst, producing propylene glycol and other lower polyols. In spite of much research effort this potentially important reaction is at present limited to the laboratory scale. One drawback is the use of high temperatures and pressures, which necessitate expensive high pressure equipment and increase the capital cost of the process. Typical hydrogen pressures between 100 and 325 bar and temperatures in the range 200–350 °C are used, and the selectivity towards propylene glycol is generally low. For example, the hydrogenolysis of glycerol under hydrogen at 300 bar and 260 °C in the presence of Raney nickel, Ru, Rh or Ir catalysts yields mainly methane, but in the presence of Raney copper 1,2-propanediol is the primary product (Figure 3.6).

Similarly, glycerol hydrogenolysis using heterogeneous metal (Cu, Pd, Rh) catalysts in a range of solvents (*e.g.*, water, sulfolane or dioxane)

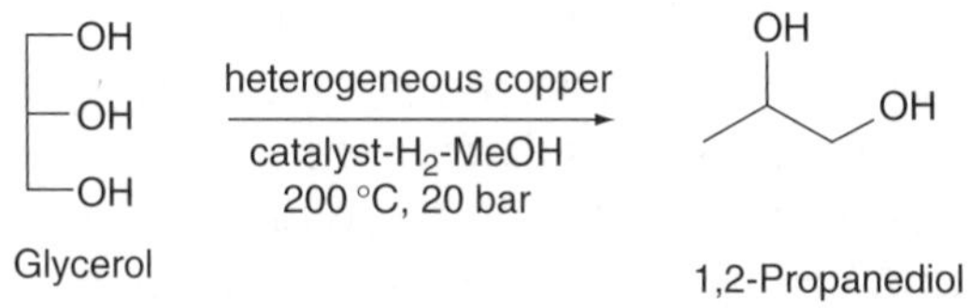

Glycerol → (heterogeneous copper catalyst-H_2-MeOH, 200 °C, 20 bar) → 1,2-Propanediol

Figure 3.5 Davy method for production of propylene glycol by glycerol hydrogenation using a copper catalyst at 200 °C and hydrogen at 20 bar.

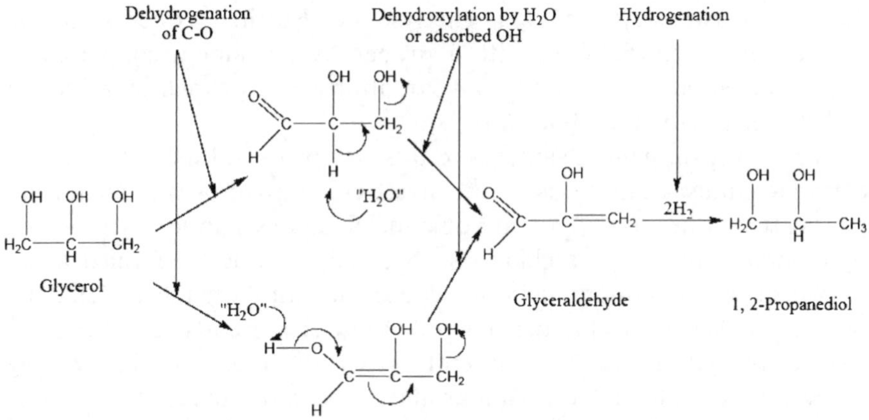

Figure 3.6 Reaction mechanism for conversion of glycerol to 1,2-propanediol over Raney copper. (Reproduced from Ref. 7, with permission.)

under hydrogen at 80 bar pressure and 180 °C, with additives to improve the reaction rate and selectivity, gives poor yields.[7] The best selectivity (100%) for 1,2-propanediol obtained by hydrogenolysis of an aqueous solution of glycerol in the presence of CuO–ZnO catalysts also gives a low yield. Similarly, hydrogenolysis of glycerol over a Ni–Re catalyst over 4 h at 230 °C under hydrogen at 82 bar produces 44% 1,2-propanediol and 5% 1,3-propanediol, together with 13% ethylene glycol.[8]

In 1985 Celanese patented the hydrogenolysis of an aqueous glycerol solution under 300 bar of syngas at 200 °C in the presence of a homogeneous rhodium complex [Rh(CO)$_2$(acac)] and tungstenic acid, giving 1,3-propanediol and 1,2-propanediol in 20% and 23% yield, respectively.[9] Shell developed the use of a homogenous palladium complex in a water–sulfolane mixture in the presence of methanesulfuric acid. After 10 h reaction, 1-propanol, 1,2-propanediol and 1,3-propanediol were present in the ratio 47:22:31.[10] A recent series of investigations into glycerol hydrogenolysis using Ru/C,[11] Rd/SiO$_2$[12] and an ion-exchange resin (Amberlyst) at 120 °C and hydrogen at 80 bar was aimed at dehydration of glycerol to acetol, catalyzed by the acidic resin and subsequent hydrogenation of acetol on the metal catalyst; this showed that neither catalyst was highly selective and a variety of products (1-propanol, 2-propanol, ethylene glycol, propylene glycol and degradation products) were eventually isolated. Finally, an additional drawback of these processes is the use of a dilute solution of glycerol for the reaction. Concentrations of glycerol as low as 10–30 wt% are predominantly used, and become further diluted by the water formed in the reaction. This reduces the average space–time yield of the reaction, increasing the energy consumption of the process and in turn decreasing its profitability.

3.3 Dehydroxylation to 1,3-Propanediol

1,3-Propanediol (PDO) is currently produced from petroleum derivatives such as ethylene oxide, using chemical catalysts developed by Shell (the "Shell route").[13] DuPont, on the other hand, has recently replaced the older acrolein (Degussa–DuPont) route[14] with a biological process starting from glucose, based on genetically engineered *Escherichia coli*.[15] The polyester formed has been desribed as the "new nylon"[16] and is used in carpet and other textile fibers, since it embodies a unique combination of chemical resistance, light stability, elastic recovery and dyeability.

In 2003 a classical synthetic approach to the production of PDO from glycerol *via* selective dehydroxylation was reported.[17] The process selectively transforms the central hydroxyl group of glycerol into a tosyloxyl group and then removes the transformed group by catalytic hydrogenolysis. The method consists of three steps: acetalization, tosylation and detosyloxylation. Compared to the hydroxyl group, the tosyloxyl group is a better leaving group and is easier to replace with a hydride ion. The first step in the conversion of glycerol to 1,3-propanediol is to acetalize the glycerol with benzaldehyde. Its purpose is protect the first and third hydroxyl groups of the glycerol so that only the central group is tosylated in the second step.

The condensation between glycerol and benzaldehyde is an equilibrium reaction, but it can be driven to completion by removing the water formed. The second step of the conversion is tosylation of the unprotected hydroxyl group of the acetalized glycerol to transform it into a good leaving group. The final step of the conversion is detosyloxylation, either preceded or followed by a hydrolysis reaction. The detosyloxylation reaction removes the tosylated central hydroxyl group, while the hydrolysis reaction deprotects the first and third hydroxyl groups. The detosyloxylation reaction shown in Figure 3.7 essentially involves hydrogenolysis. According to the proposed conversion approach, this reaction is accomplished using molecular hydrogen in the presence of a transition metal catalyst.

3.4 Biological Reduction to PDO

Glycerol can serve as a feedstock for the fermentative production of PDO. The fermentation uses bacterial strains from the groups

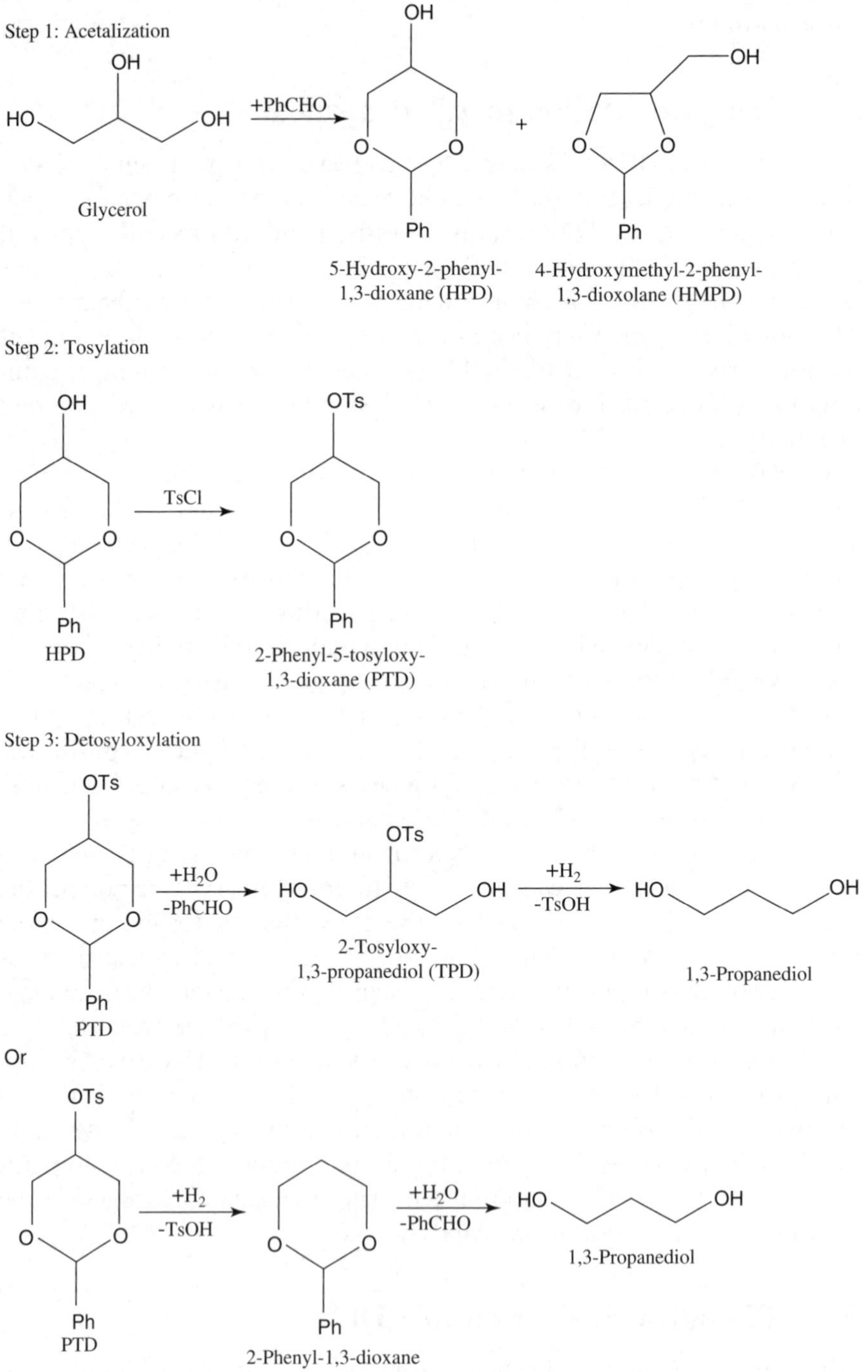

Figure 3.7 Glycerol hydrogenolysis to 1,3-PDO using a protection–deprotection approach. (Reproduced from Ref. 17, with permission.)

Citrobacter, *Enterobacter*, *Ilyobacter*, *Klebsiella*, *Lactobacillus*, *Pelobacter* and *Clostridium*. The harmless microorganism *Clostridium*, widely disseminated in nature, was shown to convert glycerol to PDO as early as 1881,[18] and has been widely investigated due to its appreciable substrate tolerance and the yield and productivity of the process. In each case glycerol is converted to PDO in a two-step enzyme-catalyzed reaction sequence. In the first step a dehydratase catalyzes the conversion of glycerol to 3-hydroxypropionaldehyde (3-HPA) and water, Equation (1). In the second step, 3-HPA is reduced to PDO by a NAD^+-linked oxidoreductase, Equation (2). The 1,3-propanediol is not metabolized further and as a result it accumulates in the medium. The overall reaction consumes a reducing equivalent in the form of a cofactor, reduced beta-nicotinamide adenine dinucleotide (NADH), which is oxidized to nicotinamide adenine dinucleotide (NAD^+), Equation (3)

$$\text{Glycerol} \rightarrow 3\text{HPA} + \text{H}_2\text{O} \tag{1}$$

$$3\text{HPA} + \text{NADH} + \text{H}^+ \rightarrow \text{PDO} + \text{NAD}^+ \tag{2}$$

$$\text{Glycerol} + \text{NAD}^+ \rightarrow \text{DHA} + \text{NADH} + \text{H}^+ \tag{3}$$

The biological process for the production of PDO has a low metabolic efficiency and uses the relatively expensive glycerol.[19] A less costly method has been developed in China starting from glucose rather than glycerol. It combines the pathway from glucose to glycerol with the bacterial route from glycerol to PDO (Figure 3.8).[20]

However, under certain conditions the classic technique based on glycerol can be attractive from both technical and economic aspects. An extensive screening of new microorganisms and use of an improved process design (fed-batch with pH-controlled substrate dosage) has allowed the product concentrations, which were relatively low at a maximum of 70 to 80 g/L as a result of product inhibition, to be increased to above 100 g/L.[21] An additional advantage of this technique and the new bacterial strains isolated is the utilization of low-priced crude glycerol or an aqueous glycerol solution. This is a factor which

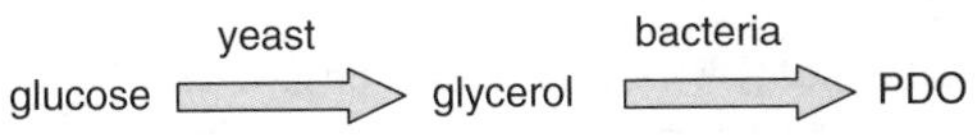

Figure 3.8 Integrated production of PDO from glucose and glycerol is being developed in China.

should not be underestimated, as it has a direct effect on the product cost. A further development is the use of immobilized rather than freely suspended cells, which enables an increase in productivity from about 2 to 30 g PDO/(L h). A comparison of existing chemical techniques with the new biotechnical approach, based on different substrates and glycerol qualities (and therefore costs), shows that biotechnology could become a competitive technique if crude glycerol was used.

3.5 Commercial Applications

The position of glycerol as a polyol meant that until the early 2000s it competed in price with other polyols available in the market (Figure 3.9).[22]

As mentioned earlier, conversion of crude glycerol to PDO has resulted in an antifreeze product comprising 70% propylene glycol and 30% glycerol, which can be produced, refined and marketed directly using existing biodiesel facilities.[23] The company, Renewable Alternatives, was the first to focus on creating an antifreeze based on glycerol mixed with propylene glycol. Indeed, propylene glycol-based antifreezes are already competing with those based on ethylene glycol, all of them approved and ready to use. Products containing propylene glycol are slightly more expensive, but the new process will bring the price down

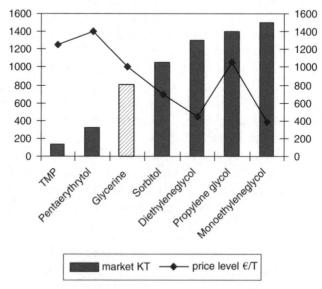

Figure 3.9   Price relationship between glycerol and other polyols in relation to market size. (Source: Novaol, May 2002; data given in kilotonnes.)

and make it the predominant product. There is also a significant positive environmental advantage, since the ethylene glycol currently in almost universal use is toxic, whereas propylene glycol is not.

However, pressing commercial considerations compelled the company to aim for production of propylene glycol in pure condition. As a result, the reactive distillation process now achieves greater than 99.8% purity, which means the product can be used both as an industrial feedstock or as an antifreeze. Indeed, the practical advantages of the reactive distillation approach are numerous:

- low water content of the feed (70–80%);
- low pressure (200 psi);
- high selectivity (>90%);
- high atom economy (>90%); and
- low-cost catalyst.

Propylene glycol is a major commodity chemical traditionally derived from propylene oxide (and hence from petroleum), with an annual global demand estimated at between 1.18 and 1.58 billion tonnes.[24] By early 2007 it was selling at around US$1.8 per kg, with a 4% annual growth in market size. However, the steeply increasing petroleum crude oil price has caused the price of propylene oxide to escalate by a factor of four over the past three years. Typical uses for propylene glycol are unsaturated polyester resins, functional fluids (antifreeze, de-icers and heat transfer liquids), pharmaceuticals, foodstuffs, cosmetics, liquid detergents, tobacco humectants, flavors and fragrances, personal care products, paints and animal feedstuffs.

The first commercial facility to use the Renewable Alternatives process is in Atlanta, where Senergy Chemical, a consortium of propylene glycol consumers and manufacturers which has licensed the process, was expected to be in full production by the end of 2007.

Another glycerol hydrogenolysis process soon to be commercialized has been developed in the UK by Davy, a Johnson Matthey company (Figure 3.10). Davy is using its new glycerol to propylene glycol process in a joint venture with Ashland–Cargill. The new company will produce in excess of 65 000 tonnes of propylene glycol per annum at a European location, starting early in 2009.[25] Indications suggest that the process will give high purity propylene glycol, suitable for all applications.

In the Davy process glycerol is reacted with hydrogen over a heterogeneous copper catalyst under relatively moderate conditions (20 bar, 200 °C). The glycerol, along with a recycle stream, is vaporised in a recirculating stream of hydrogen, typically from a pressure-swing

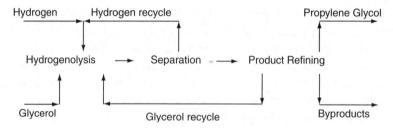

Figure 3.10   Glycerol hydrogenolysis process to propylene glycol developed by Davy
in the UK. (Reproduced from Ref. 25, with permission.)

adsorption unit. Glycerol conversion is around 99% and by-products
are removed by distillation. The advantage of the Davy process is its
high selectivity to the desired product. The refining process shown in
Figure 3.10 delivers propylene glycol at high purity, and the water
produced in the reaction is of sufficient quality for biological treatment.
The propylene glycol meets the specification required for producing
unsaturated polyester resins and functional fluids, and pharmaceutical
grade material can also be produced if required. The relatively small by-
product streams are of high quality and can be used as solvents or, in the
case of mixed glycols, in functional fluids. Finally, the Dow Haltermann
Custom Processing company has contracted to convert glycerol to
propylene glycol at a facility in Houston. The company already had
limited commercial quantities available by mid-2007.

References

1. S. Howell, US National Biodiesel Board, cited in: D. Nilles, Combating the glycerin glut, *Biodiesel Magazine*, September 2006.
2. For updated information, visit the website of Senergy Chemical Inc
   at http://www.senergychem.com.
3. M. Dasari, P. Kiatsimkul, W. Sutterlin, G. J. Suppes, Low-pressure
   hydrogenolysis of glycerol to propylene glycol, *Appl. Catal. A*, 2005,
   281, 225. The process has been patented: G. J. Suppes, W. R.
   Sutterlin, M. A. Dasari, Method for producing lower alcohols from
   glycerol, WO2005095536.
4. G. J. Suppes, Catalysis and methods for improved glycerol reduction, Presentation at Bio-Futures Conference: 16–17 October,
   2006, Saskatoon (Canada).
5. C. W. Chiu, M. A. Dasari, W. R. Sutterlin and G. J. Suppes, Removal of residual catalyst from simulated biodiesel crude glycerol
   for glycerol hydrogenolysis to propylene glycol, *Ind. Eng. Chem.
   Res.*, 2006, **45**, 791.

6. M. W. M. Tuck and S. N. Tilley, Vapor-phase hydrogenation of glycerol, WO2007010299.

7. J. Chaminand, L. Djakovitch, P. Gallezot, P. Marion, C. Pinel and C. Rosier, Glycerol hydrogenolysis on heterogeneous catalysts, *Green Chem.*, 2004, **6**, 359.

8. T. Werpy, J. Frye, A. Zacher and D. Miller, Hydrogenolysis of 6-carbon sugars and other organic compounds using multimetallic catalysts, WO03035582 (2002).

9. T. M. Che, Catalytic conversion of glycerol and synthesis gas to propanediols, US4642394 (1987).

10. E. Drent, W. W. Jager, Process and catalysts for the hydrogenolysis of glycerol into 1,3-propanediol and acrolein, US6080898 (2000).

11. For example, Ru/C coupled to an acid catalyst such as the ion-exchange resin Amberlyst, which is effective in the hydrogenolysis of glycerol under mild reaction conditions (393 K, 8.0 MPa): T. Miyazawa, Y. Kusunoki, K. Kunimori, K. Tomishige, *J. Catal.*, 2006, **240**, 213.

12. I. Furikado, T. Miyazawa, S. Koso, A. Shimao, K. Kunimori and K. Tomishige, Catalytic performance of Rh/SiO_2 in glycerol reaction under hydrogen, *Green Chem.*, 2007, **9**, 582.

13. K. T. Lam, J. P. Powell and P. R. Wieder, Preparing 1,3-propanediol, WO9716250 (1997).

14. D. Arntz, T. Haas, A. Muller and N. Wiegand, Process for the production of 3-hydroxyalkanals, US5276201 (1994).

15. Namely, *E. coli* transformed with the *Klebsiella pneumoniae* dihydroxyacetone regulon genes. The process relies on the presence in *E. coli* of a gene encoding a non-specific catalytic activity sufficient to convert 3-hydroxypropionaldehyde to 1,3-propanediol: M. Emptage, S. L. Haynie, L. A. Laffend, J. P. Pucci, G. M. Whited, Process for the biological production of 1,3-propanediol with high titer, US7067300 (2006).

16. B. Balmer, PTT-A new alternative for fibres and textiles? Report by Frost & Sullivan (2004).

17. K. Wang, M. C. Hawley and S. J. DeAthos, Conversion of glycerol to 1,3-propanediol via selective dehydroxylation, *Ind. Eng. Chem. Res.*, 2003, **42**, 2913.

18. R. Lin, H. Liu, J. Hao, K. Cheng and D. Liu, *Biotechnol. Lett.*, 2005, **27**, 1755.

19. H. Biebl, K. Menzel, A.-P. Zeng and W.-D. Deckwer, *Appl. Microbiol Biotechnol.*, 1999, **52**, 289; H. Petitdemange, H. Biebl, Production of 1,3-propanediol from glycerol surpluses: Yield

optimisation by technological development and by genetic strain improvement, http://www.biomatnet.org/secure/Air/S233.htm.

20. D. Liu, Integrated production for biodiesel and 1,3-propanediol with lipase-catalyzed transesterification and fermentation. Presentation at the Japanese Institute of Energy, 2006: http://www.jie.or.jp/pdf/ 21%5B1%5D.Prof.DehuaLIU.pdf#search=%22Liu%2Bpropanediol %22.

21. S. Hirschmann, K. Baganz, I. Koschik and K.-D. Vorlop, Development of an integrated bioconversion process for the production of 1,3-propanediol from raw glycerol waters. *Landbauforsch Völkenrode*, 2005, **55**, 261.

22. For instance, NOF Corporation in Japan, has recently developed a new antifreeze, Camag, composed of glycerol and potassium acetate, to prevent freezing of roads in the cold northern areas of Japan.

23. For this achievement, Professor G. J. Suppes has been awarded the 2006 Presidential Green Chemistry Challenge Awards: http:// www.epa.gov/greenchemistry/pubs/pgcc/winners/aa06.html; Polyol Partners can also hydrocrack glycerol and form propylene glycol. C. Boswell, *Chemical Marketing Reporter*, 24 January, 2005.

24. S. Shelley, A renewable route to propylene glycerol. *Chem. Eng. Prog.*, 2007, **103**(8), 6.

25. See also the company's website: www.davyprotech.com.

CHAPTER 4
Halogenation

4.1 Chlorination of Glycerol

Studies of glycerol halogenation focus on the formation of 1,3-dichloro-2-propanol, an intermediate in epichlorohydrin synthesis, by direct hydrochlorination (Figure 4.1).

1,3-Dichloro-2-propanol (α,γ-chlorohydrin) is the isomer of choice because epichlorohydrin, formed by its dehydrochlorination, is so important commercially. By condensation with a polyol such as bisphenol A epichlorohydrin enables a linear structure to be conserved in the ether polymers obtained from it, including the commercially valuable epoxy resins (Figure 4.2).

Liquid epoxy resins are used in marine protective, automotive, housing and can coatings, and many other applications.[1] Epoxy resins are used with over 400 000 tonnes of curing agents annually to produce an estimated 3 million tonnes of products, worth over US$20 billion.[2]

Epichlorohydrin is traditionally derived indirectly by reacting propylene with chlorine. Annual production is around 903 000 tonnes. The industrial process (Figure 4.3) yields a mixture of 1,2-dichloropropanol (30%) and 1,3-dichloropropanol (70%). The 1,2-dichloropropanol requires further conversion to the 1,3-isomer, slowing down the process. The overall yield is 97% but large amounts of waste water are produced, contaminated with chlorinated by-products such as 1,2- and 1,3-dichloropropane, 1,2,3-trichloropropane, penta- and hexachlorohexanes, in addition to 1,3- and 1,2-dichloropropanol. For instance, the world's largest manufacturer, Dow Chemical, produces 45 000 tonnes of waste water annually using the traditional process involving high-temperature chlorination of propene followed by hydrolysis.[2]

RSC Green Chemistry Book Series
The Future of Glycerol: New Uses of a Versatile Raw Material
By Mario Pagliaro and Michele Rossi
© Mario Pagliaro and Michele Rossi 2008

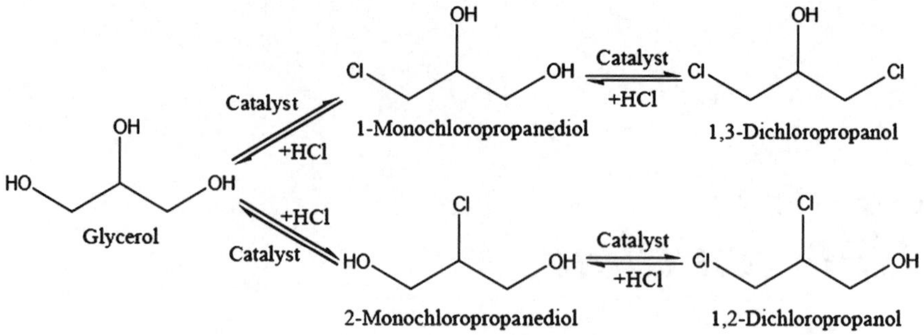

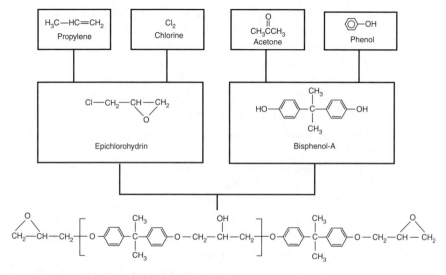

Figure 4.1   Non-selective hydrochlorination of glycerol yields 1,3-dichloro-2-propanol and 1,2-dichloro-3-propanol. Water is a by-product of the reaction.

Figure 4.2   Manufacturing process for bisphenol A-based epoxy resins.

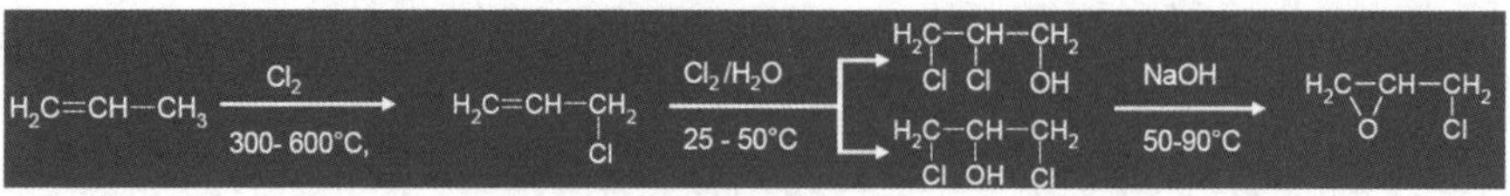

Figure 4.3   Traditional industrial process for epichlorohydrin ("Epi" in industry jargon) based on high temperature chlorination of propene.

Described as early as 1931, the traditional process for hydrochlorination of glycerol has poor selectivity and very low productivity. It involves reaction of glycerol with aqueous hydrochloric acid in the presence of acetic acid as catalyst at a temperature of 80–100 °C. Initial chlorination gives primarily 1-monochloropropanediol and water, along with small quantities of 2-monochloropropanediol. This is followed by a second chlorination from which 1,3-dichloropropanol is obtained, with modest quantities of 1,2-dichloropropanol as by-product. A number of later patents describe biphasic processes using water and organic solvent, in which 1,3-dichloropropanol is rendered soluble by performing the reaction at the boiling point of the mixture; the actual temperature varies depending on the particular solvent used.[3] All the traditional processes starting from glycerol have considerable drawbacks, including the loss of catalyst during the reaction due to the low boiling point of acetic acid, slowing of the reaction caused by the introduction of water into the reaction mixture in the form of aqueous HCl, and difficulty in recovering the α,γ-chlorohydrin from the reaction mixture.

4.2 Production of Epichlorohydrin

A more recent hydrochlorination process allows the direct synthesis of 1,3-dichloropropanol from glycerol and hydrochloric acid, and dehydrochlorination using sodium hydroxide generates epichlorohydrin (Figure 4.4).

A modified glycerol-based selective process for the production of 1,3-dichloro-2-propanol from either pure or raw glycerol (Figure 4.5) uses

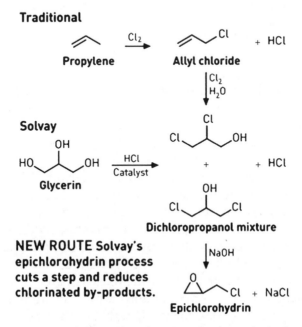

Figure 4.4 Traditional and glycerol-based processes for the synthesis of epichlorohydrin.

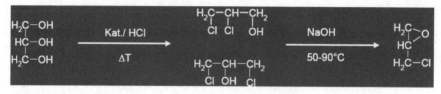

Figure 4.5 The glycerol-based process for epichlorohydrin.

gaseous anhydrous hydrogen chloride in the presence of a low volatility organic acid catalyst. It also employs a system for the continuous removal of the reaction water to improve the efficiency and economy of the process.[4]

The use of hydrogen chloride in the gaseous phase rather than in aqueous solution avoids the introduction of water, which has a negative effect on the reaction balance. The use of catalysts based on carboxylic acids having 3 to 10 carbon atoms, with boiling points above 120 °C, as an alternative to acetic acid compensates for the loss of catalyst due to the reaction temperature, which is close to the boiling point of acetic acid; the concentration of catalyst thus remains constant during the reaction. The α,γ-dichlorohydrin is recovered together with the water of reaction, for example by performing the reaction in a continuous stream of hydrogen chloride with stripping, since α,γ-dichlorohydrin is much more volatile than glycerol or monochlorohydrins. The reaction temperature is controlled between 80 and 180 °C and the pressure of gaseous hydrogen chloride is below 5 bar. The hydrogen chloride pressure has a positive effect on the reaction rate but adversely affects the stripping of the products; a reasonable compromise is to operate within the range 1 to 5 bar. As an example, anhydrous glycerol and malonic acid (8 mol% based on glycerol) are loaded into a reactor and the temperature increased to 100 °C. Under normal conditions a stream of hydrogen chloride gas is introduced at the rate of 50 L h^{-1}. As the reaction proceeds the products are removed, and after 1 h the glycerol is completely converted. The desired dichlorohydrin isomers are obtained in 76.5% yield, the remainder being predominantly monochlorohydrin (Figure 4.6).

The three-step reaction mechanism shown in Figure 4.7 involves a rate-determining esterification step *via* nucleophilic substitution on the acylic carbon with formation of water (**1**), followed by the formation of an oxonium group through alkyl–oxygen bond scission and carboxylic acid release (**2**), and the subsequent formation of chlorohydrin by chloride addition to either the α- or the β-carbon atom (**3**).[5]

This reaction mechanism explains why the formation of α-monochlorohydrin is always higher than that of the β-isomer. Since the latter species cannot react further its concentration increases slightly

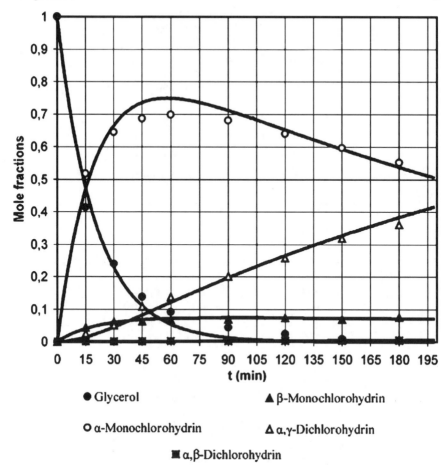

Figure 4.6 Evolution of the composition of a reaction mixture of glycerol and HCl with malonic acid as catalyst at 120 °C. Comparison between the experimental data and the model behavior is also shown. (Reproduced from Ref. 5, with permission.)

during the reaction until all the glycerol has been consumed, and it then remains almost constant. On the other hand α-monochlorohydrin is able to undergo a second chlorination with a similar mechanism, affording α,γ-dichlorohydrin (Figure 4.8).

The entry of the first chlorine atom in α-monochlorohydrin favors the attack of the second chlorine in the γ-position due to an induction effect. Calculations show that substitution in the γ-position is 7.4 times slower than the first chlorination, whereas attack of chlorine on the carbon in the β-position is 510 times slower than the first chlorination.

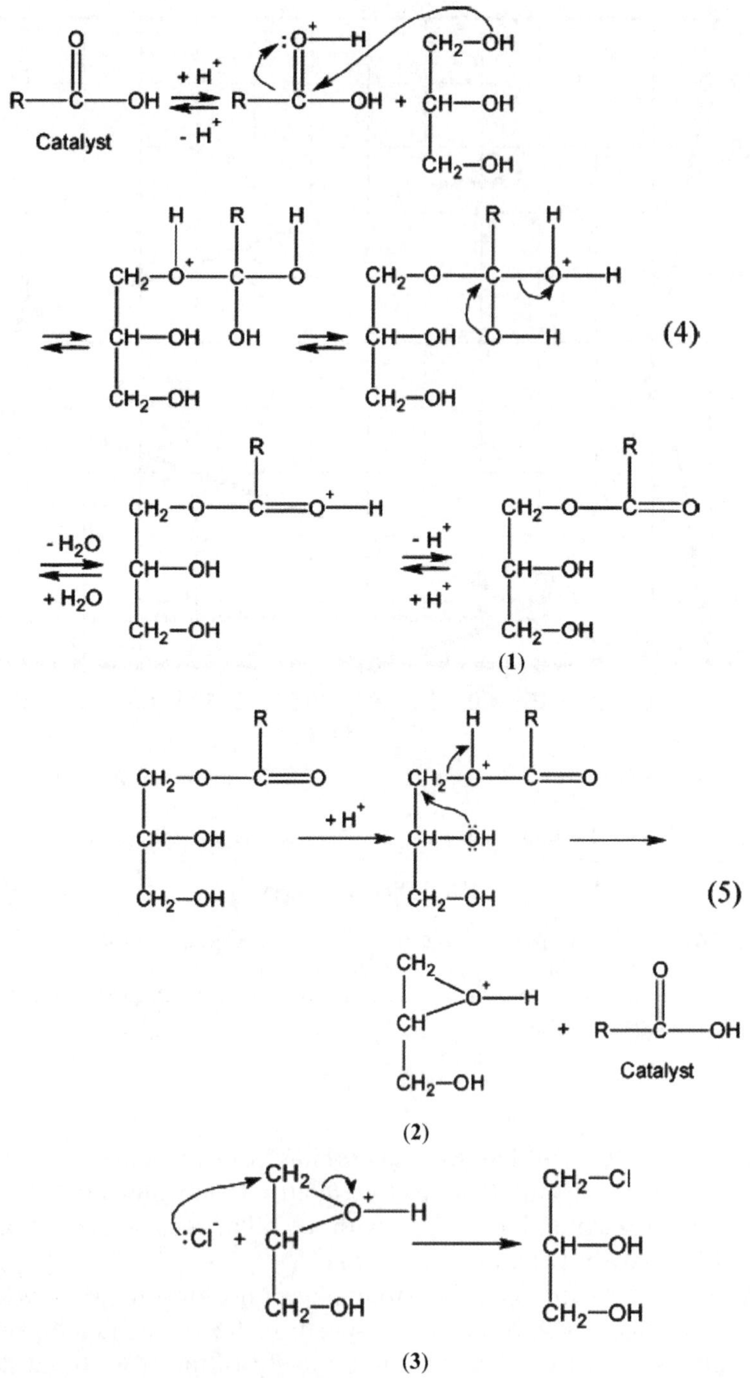

Figure 4.7 The three-step reaction mechanism for glycerol chlorohydrin formation *via* hydrochlorination, explaining the observed reaction path. (Reproduced from Ref. 5, with permission.)

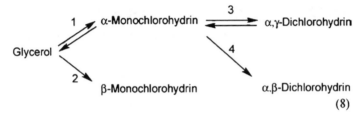

Figure 4.8 Reaction path for hydrochlorination of glycerol. (Reproduced from Ref. 5, with permission.)

4.3 Industrial Applications

In 2007 Solvay, a traditional glycerol and epichlorohydrin manufacturer, was the first to start production of epichlorohydrin from glycerol, at their 10 000 tonne plant in France. Glycerol was obtained from a French source as a by-product of the manufacture of biodiesel from rapeseed oil. In the early 2000s Solvay was facing an economically unsustainable situation in the production of chlorinated organics due to soaring propylene prices, which had increased by a factor of four since 1999 with a concomitant fall in the price of glycerol. The company therefore halted production of synthetic glycerol from epichlorohydrin in 2005, aiming to reverse to procedure by converting the plant to produce epichlorohydrin from glycerol (Figure 4.9).

The proprietary process is again based on organic acid catalysts. Optimal reaction occurs using anhydrous hydrochloric with 30 mol% caprylic acid as catalyst at above 120 °C, which ensures that only a limited fraction (10%) of the catalyst evaporates from the reactor.[6] In order to avoid corrosion of the glass-lined steel reaction vessel, the manufacture of dichloropropanol is carried out keeping the inner wall of the vessel which lies above the level of the liquid medium at a temperature of 120 °C, at which corrosion of the enameled steel is minimized.[7] In the subsequent step, which produces epichlorohydrin by dechlorination with NaOH, the aqueous fraction rich in NaCl can be recovered and used in the production of chlorine by electrolysis. The fraction rich in water is recycled to provide the aqueous solution needed for hydrodechlorination. The new glycerol-based process shows crucial advantages over the existing propene route:

- it does not require a solvent;
- the size of the reactors can be reduced thanks to higher selectivity;
- the kinetics is much faster;
- hydrogen chloride is consumed rather than produced;
- chlorine consumption is reduced by 50% and water by 70%; and
- chlorinated residues are 80% lower.

Epicerol™, from product to raw material

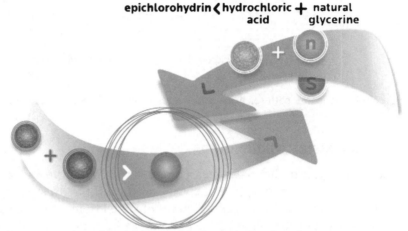

epichlorohydrin ❰ **hydrochloric** ➕ **natural**
acid   **glycerine**

chlorine ➕ **propylene** ❱ **epichlorohydrin** ❱ **synthetic glycerine**

Figure 4.9   Hydrochlorination of glycerol actually inverts the traditional manufacturing process for glycerol. (Reproduced from the Solvay website, with
permission.)

Moreover, the process can run either batch-wise or continuously. Increasing demand for epichlorohydrin is expected to exceed the existing
global production capacity by 2010, and the company is already planning
further investment in a 100 kilotonne unit on its integrated site at Map Ta
Phut, Thailand, where production will begin mid-2009. In addition, Dow
has announced the construction of a large glycerol-to-epichlorohydrin
plant in China, starting production in 2010. The company has selected the
Shanghai Chemical Industry Park for its 150 kilotonne plant.[8] In this
case, glycerol will be purchased from local producers of biofuels, which in
China are typically obtained from rapeseed and palm oil. Dow has also
decided to build a 100 kilotonne liquid epoxy resin plant at the Shanghai
location. The Dow production facility will reduce waste water by more
than 70% compared to conventional propylene-based technology and will
almost completely avoid the formation of organic by-products.

Again, the glycerol-to-epichlorohydrin process[9] is based on organic
acid-catalyzed hydrochlorination and offers similar advantages to those
listed above. The conversion process from crude glycerol uses glacial
acetic acid and anhydrous hydrogen chloride at 8.2 bar at 120 °C for
90 min. The procedure takes place twice consecutively and gives a
mixture of l,3-dichloropropan-2-ol (95.3 wt%), 2,3-dichloropropan-l-ol
(2.6 wt%), with trace amounts of 2-acetoxy-1,3-dichloropropane
(0.7 wt%) and 1-acetoxy-2,3-dichloropropane (0.1 wt%). Importantly,

none of the highly toxic trichloropropanol is formed, and the glycerol is converted completely.

References

1. Epichlorohydrin is also employed for water purification and as a reinforcement agent for paper, for instance in the food industry in the manufacture of tea bags. In addition, epichlorohydrin is used in the pharmaceutical industry as a synthetic starting point and as a base for synthesis of glycerol monochlorohydrin (1-chloro-2,3-propanediol), employed in the manufacture of pharmaceutical products such as cough mixtures. *Ullmann's Encylopedia of Industrial Chemistry*, 7th edn, Weinheim: Wiley–VCH, 2006.
2. R. Busch, Renewable raw materials for the chemical industry, NATO ASI "New organic chemistry reactions and methodologies for green productions", 30 October – 5 November 2006, Lecce, Italy.
3. (a) F. N. Grimsby, Dichlorohydrin, US4620912, 1986; (b) Y. Morozov, Glycerol 1,3-dichlorohydrin, RU1225836 (1986); (c) C. Divo, M. Petri, M. Lazzari, A. Bigozzi, Process for the preparation of glycerol dichlorohydrins, GB2029821 (1980).
4. D. Siano, E. Santacesaria, V. Fiandra, R. Tesser, G. Di Nuzzi, M. Di Serio, M. Nastasi, Continuous regioselective process for the production of 1,3-dichloro-2-propanol from glycerol and hydrochloric acid in the presence of organic carboxylic acid catalysts, WO2006111810.
5. R. Tesser, E. Santacesaria, M. Di Serio, G. Di Nuzzi, V. Fiandra, Kinetics of glycerol chlorination with hydrochloric acid: A new route to α, γ-dichlorohydrin, *Ind. Eng. Chem. Res.*, 2007, **46**, 6456.
6. P. Krafft, P. Gilbau, B. Gosselin, S. Claessens, Process for producing dichloropropanol from glycerol, the glycerol eventually from the conversion of animal fats in the manufacture of biodiesel, WO2005054167.
7. P. Krafft, C. Franck, I. De Andolenko, R. Veyrac, Process for the manufacture of dichloropropanol by chlorination of glycerol, WO2007054505.
8. J.-F. Tremblay, Dow selects Shanghai for epichlorohydrin. *Chem. Eng. News* (March 29, 2007).
9. D. Schreck, W. J. Kruper, R. D. Varjian, M. E. Jones, R. M. Campbell, K. Kearns, B. D. Hook, J. R. Briggs, J. G. Hippler, Conversion of multihydroxylated aliphatic hydrocarbon or ester thereof to a chlorohydrin, WO2006020234.

CHAPTER 5
Dehydration

5.1 Dehydration of Glycerol

Two important chemicals can be produced directly by dehydration of glycerol: acrolein and 3-hydroxypropionaldehyde (3-HPA). In addition, oxydehydration of glycerol gives the commercially important acrylic acid.

5.2 Dehydration to Acrolein

Acrolein is an important and versatile intermediate for the chemical industry, which can be obtained by dehydration of glycerol (Figure 5.1).[1]

Acrolein can be polymerized into acrylic resins and forms the basis of superabsorbent polymers, widely employed in the baby hygiene market (Figure 5.2).

When glycerol is protonated the energy barrier for dehydration is greatly reduced, falling from about 60 to $20\,kJ\,mol^{-1}$ (Figure 5.3).[2] For this reason, dehydration to acrolein is carried out under acidic conditions.

Acrolein is an explosive (Figure 5.4) and toxic chemical whose handling requires the highest safety standards. The most significant direct application of acrolein is as a herbicide for the control of aquatic plants. It kills plant cells by the destruction of cell membrane integrity, and also by its affinity for sulfhydryl groups, causing the denaturation of vital enzymes. Furthermore, it is used in the manufacture of 3-methylthio-propionaldehyde (known as MMP), a precursor (by reaction with HCN) of methionine, an essential amino acid required by animals to meet the nutritional needs for proper growth, health and reproduction.

As a general rule the hydration reaction is favored at low temperatures and the dehydration reaction at high temperatures. To obtain acrolein it is thus necessary to use a sufficiently high temperature, and/or

RSC Green Chemistry Book Series
The Future of Glycerol: New Uses of a Versatile Raw Material
By Mario Pagliaro and Michele Rossi
© Mario Pagliaro and Michele Rossi 2008

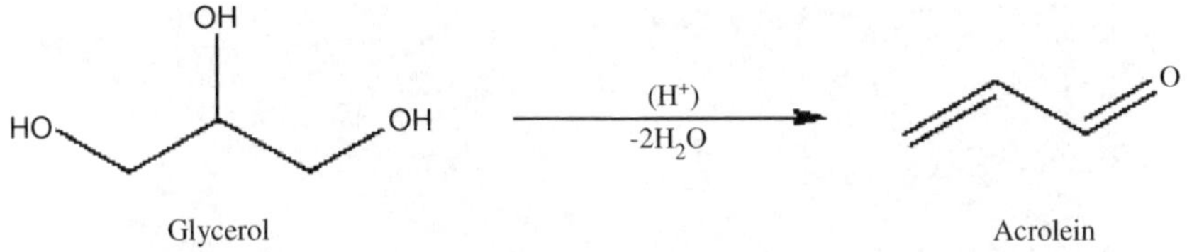

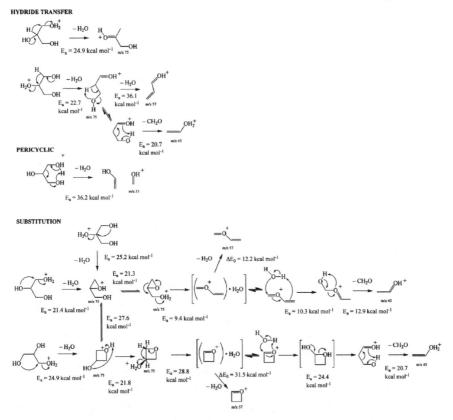

Figure 5.1 Acid-induced dehydration of glycerol to acrolein proceeds smoothly at 250–340 °C over a heterogeneous acidic catalyst.

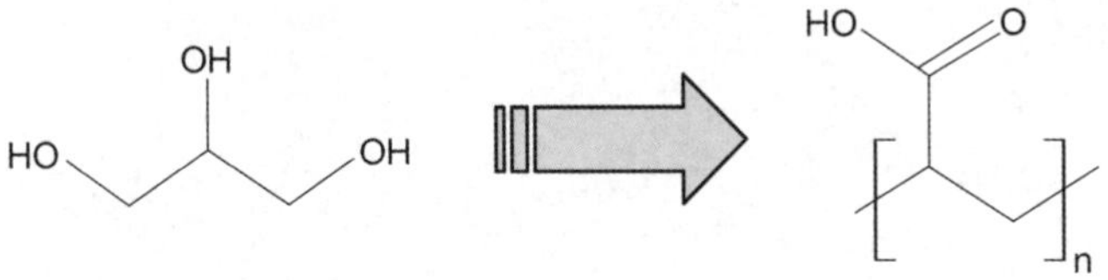

Figure 5.2 Acrolein produced from glycerol is further oxidized to acrylic acid, which can in turn be polymerized to poly(acrylic acid).

Figure 5.3 Mechanisms of acid catalyzed dehydration of glycerol as suggested by quantum mechanics calculations. (Reproduced from Ref. 2, with permission.)

Figure 5.4   Acrolein, a deadly toxin, is also a powerful explosive. This photograph
from 11 December 1982 shows the consequences of a tank containing
acrolein exploding at a Taft, Louisiana, plant. Windows a mile and a half
from the plant were blown out, and 17 000 people were evacuated.
(Photo courtesy of Isaac CC on Flickr.com.)

partial vacuum, to shift the reaction. The reaction may be performed in
the liquid or the gas phase and it is known to be catalysed by acids.

In the mid-1990s Degussa introduced a method based on the dehydration of glycerol on solid acidic catalysts.[3] For example, reacting a
gaseous glycerol–water mixture, with a glycerol content of 10 to 40 wt%,
at 250–340 °C on a solid acidic catalyst with a Hammett acidity function
(H_0 value) below 2, results in complete conversion of the glycerol to
acrolein. In this process, however, the dehydration of glycerol to acrolein is generally accompanied by side-reactions giving by-products such
as hydroxypropanone, propanaldehyde, acetaldehyde, acetone, adducts
of acrolein with glycerol, glycerol polycondensation products and cyclic
glycerol ethers, which can cause the formation of coke on the catalyst.
This results in deactivation of the catalyst, reduction in yield and in the
selectivity towards acrolein. The presence of by-products in the acrolein
such as hydroxypropanone or propanaldehyde, which can be difficult
to isolate, necessitates separation and purification steps which lead to a
high recovery cost for the purified acrolein. Moreover, it is necessary
to regenerate the catalyst regularly in order to maintain its activity. As a
result the process has not been commercialized, and the oxidation
of propylene with a Bi–Mo mixed oxide catalyst is more economically

attractive. In practice, the technique for manufacturing acrolein is normally based on the oxidation of propylene derived from petroleum.[4]

An improved process for manufacturing acrolein uses the gas-phase dehydration of glycerol using solid catalysts with H_0 between -9 and -18.[5] Such catalysts are more active and are less quickly deactivated, permitting longer cycles and a smaller reactor volume. In a typical reaction a reactor consisting of a tube 85 cm long, with an interior diameter of 6 mm, is used to perform the glycerol dehydration in the gas phase at atmospheric pressure. The reactor, containing a ground solid tungstated zirconia (ZrO_2–WO_3) catalyst (particle size 0.5 to 1.0 mm), is placed in a heated chamber maintained at 300 °C. Tungstated zirconia oxides of this type are more active than sulfated oxides, and are in general also more stable.[6]

The reactor is fed with an aqueous solution containing 20% by weight of glycerol (flow rate, 12 mL h^{-1}). After 7 h, corresponding to the passage of about 80 mL of aqueous glycerol solution over the catalyst, 84% of the glycerol has been converted and acrolein is obtained in 35% yield, along with smaller quantities of hydroxypropanone (14%), acetaldehyde (3.4%) and propanaldehyde (7.8%).

By carrying out the reaction in the presence of oxygen it is possible to obtain higher glycerol conversion by inhibiting, firstly, the formation of by-products, particularly those originating from the hydrogenation of dehydrated products such as propanaldehyde and hydroxypropanone and, secondly, the deactivation of the catalyst by reducing the formation of coke.[7] The proportion of oxygen is chosen to lie outside the flammability range at all stages; in practice this means not allowing the proportion of oxygen to exceed 7% of the gases involved in the reaction, which involve a mixture of glycerol, water, oxygen and inert gases.

In a typical conversion the tubular reactor containing the solid tungstated zirconia catalyst is placed in a heated chamber at 300 °C. The reactor is fed with a 20 wt% aqueous glycerol solution at a flow rate of 12 mL h^{-1} and oxygen at 0.8 L h^{-1}. In this instance the relative proportions of oxygen : vaporized glycerol : steam are 6.0 : 4.5 : 89.5. The aqueous glycerol solution is vaporized in the heated chamber and then passes over the catalyst, the calculated contact time being about 2.9 s. After 7 h, which corresponds to about 80 mL of aqueous glycerol solution passing over the catalyst, all the glycerol has been converted and acrolein is obtained in 53% yield, along with by-products such as acetaldehyde (10%) and propanaldehyde (*ca.* 4%). In general, the most selective catalysts for the dehydration reaction have H_0 in the range -8.2 to -3.0.[8] Loss of water from glycerol through substitution results in either oxirane or oxetane intermediates, which interconvert with a low

energy barrier. Subsequent decomposition of these intermediates proceeds either by a second dehydration step or by loss of formaldehyde.

Dehydration of glycerol also occurs in the gas phase over silica-
supported heteropolyacids. Silicotungstic acid, $H_4SiW_{12}O_{40}.24H_2O$ (HSiW)
supported on silica with mesopores of 10 nm displays the highest catalytic
activity, with acrolein selectivity at 275 °C and ambient pressure greater
than 85 mol% at complete conversion.[9] Silica Q10, having the largest mesopores, is the most appropriate support for the reaction (Figure 5.5).

Deactivation is caused by the deposition of coke on the support: the
color of the catalysts changes from white to black, with a weight increase
of the Q10–SiW–30 catalyst of 7.8% after 5 h. This discovery is of
fundamental practical importance, since supported Keggin-type hetero-
polyacid catalysts display high thermal stability, are less harmful to the
environment than mineral acids and are increasingly employed industrially for reactions such as the dehydration of 2-propanol.[10]

The reaction route (Figure 5.6) involves individual reaction mechanisms revealed by quantum mechanical computation.[2] When protonation occurs at the secondary hydroxyl group of glycerol a water
molecule and a proton are eliminated from the protonated molecule,
and 3-hydroxypropanal is then formed by tautomerism.

In practice 3-hydroxypropanal is not detected, however, since this
compound is unstable and is readily dehydrated to acrolein. In contrast,
when protonation occurs at a terminal hydroxyl group in glycerol,
hydroxyacetone is produced through dehydration and deprotonation
accompanied by tautomerism.

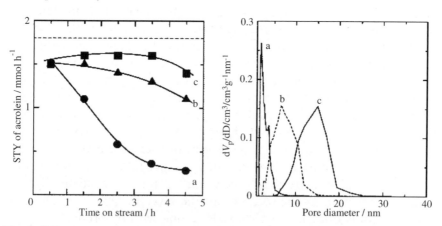

Figure 5.5   Changes (*left*) in space–time yield of acrolein in the dehydration of glycerol
at 275 °C catalyzed by silicotungstic acid supported on silica of three
different pore sizes (*right*), (a) Q3–SiW–30, (b) Q6–SiW–30, and (c) Q10–
SiW–30. The broken line indicates the maximum theoretical yield.
(Reproduced from Ref. 9, with permission.)

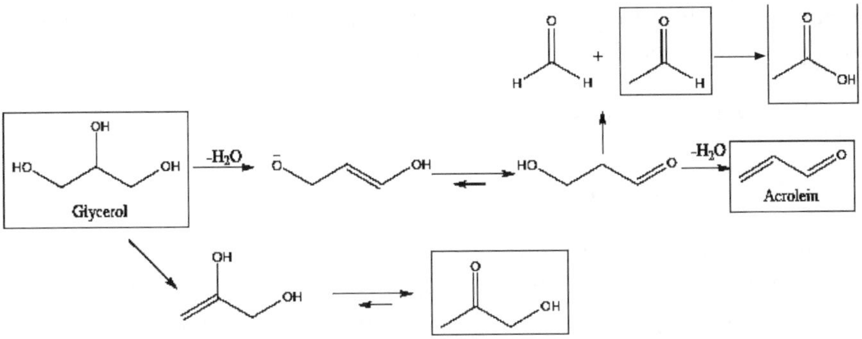

Figure 5.6 Glycerol dehydration routes, involving the individual reaction mechanisms revealed by quantum mechanical computation. (Reproduced from Ref. 2, with permission.)

Another synthetic approach is based on reaction in sub- and super-critical water as the reaction medium; this follows from the discovery that acrolein can be obtained with 84% selectivity and 40% conversion from glycerol by adding 5 mM H_2SO_4 to water at 623 K and 34.5 MPa.[11] In an improved version of this method, the reaction catalyzed by zinc sulfate in sub- and supercritical water (573–663 K, 25–34 MPa, 10–60 s) affords 75% maximum selectivity of acrolein at 50% glycerol conversion.[12] Finally, approximately 80% selectivity of acrolein at 90% glycerol conversion has been obtained by carrying out the reaction in pressurized hot water under supercritical conditions (673 K and 34.5 MPa) using H_2SO_4 as catalyst.[13] The yield of acrolein can be enhanced by increased concentrations of glycerol and H_2SO_4, and by higher pressures.

A detailed kinetic model for analyzing the dehydration reaction in near- and supercritical water at 622–748 K and 25–45 MPa without additives has concluded that conversion of glycerol progresses competitively through both ionic and radical mechanisms.[14] The respective predominance of the ionic or the radical pathway can be controlled by temperature and pressure: the ionic reaction is preferred in subcritical water (liquid state), whereas the radical mechanism is favored in the supercritical region. This kinetic model also suggests that dehydration of glycerol to acrolein occurs mainly *via* ionic reactions, whereas the competitive dehydration into allyl alcohol and bond scission to acetaldehyde and formaldehyde preferentially follows a radical mechanism (see Figure 5.6).

5.3 Oxydehydration to Acrylic Acid

Acrylic acid can be produced by a one-step oxydehydration reaction of glycerol in the presence of molecular oxygen.[15] The dehydration reaction is followed by aerobic oxidation of the resulting acrolein directly to acrylic acid, Equations (1) and (2):

$$CH_2OH-CHOH-CH_2OH \rightarrow CH_2=CH-CHO + 2H_2O \qquad (1)$$

$$CH_2=CH-CHO + 1/2O_2 \rightarrow CH_2=CH-COOH \qquad (2)$$

The exothermic oxidation reaction is thus coupled with an endothermic dehydration reaction, which provides a better thermal balance for the overall process. In a typical conversion the tubular reactor comprises a tube 85 cm long with inside diameter 6 mm, and this is used to perform the glycerol oxydehydration reaction in the gas phase at atmospheric pressure. The reactor contains two catalytic beds, the first with a ground solid tungstated zirconia catalyst, and the second with an industrial mixed W–Sr–V–Cu–Mo oxidation catalyst with acetic acid as binder. The gas mixture fluxes consecutively through the first and the second beds. The reactor is placed in a heated chamber at 280 °C and is fed with an aqueous solution containing 20 wt% of glycerol (flow rate, $9\,g\,h^{-1}$) and oxygen (flow rate, $14\,mL\,min^{-1}$). The time of contact is about 2.9 s, after which sampling confirms that all the glycerol has been converted. Acryclic acid is obtained in 74% yield.

5.4   Dehydration to 3-Hydroxypropionaldehyde

3-Hydroxypropionaldehyde (3-HPA) is a precursor for many modern chemicals, including acrolein, acrylic acid and 1,3-propanediol, and is also used for polymer production. It can be efficiently produced from renewable glycerol. Biotechnological production has several advantages compared to chemical methods and can be carried out using aqueous glycerol solution, either at room temperature or at 37 °C under normal pressure.

The transformation to 3-HPA is a one-step enzymatic reaction (Figure 5.7), and yields of 85–87 mol% 3-HPA per mol glycerol are higher than those achieved by chemical synthesis. To date, a number of genera of bacteria have been identified as being suitable, including *Klebsiella* (*Aerobacter*), *Citrobacter*,[16] *Enterobacter*,[17] *Clostridium* and *Lactobacillus*.[18] The enzyme responsible for the formation of 3-HPA from glycerol, glycerol dehydratase, is known to be cobamide-dependent.

5.5   Industrial Applications

Acrolein is currently produced by the oxidation of propylene. However, as mentioned in Chapter 4, the price of propylene has increased by a

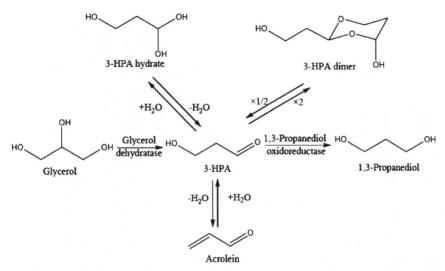

Figure 5.7 Formation of 3-hydroxypropionaldehyde (3-HPA), 1,3-propanediol and acrolein.

factor of four since 1999, making the production of acrolein from the less expensive glycerol commercially attractive. Indeed, the first facility for producing acrolein from glycerol is soon expected to be established by Arkema at Beaumont, Texas (Figure 5.8). This is a hybrid process in which the oxidation of propylene to acrolein is coupled with the dehydration of glycerol. The first glycerol dehydration step is performed in the presence of the reaction gas originating from the oxidation of propylene to acrolein.[19]

This has the advantage that the endothermic dehydration reaction is coupled to the exothermic oxidation of propene, with considerable energy savings, in contrast to the traditional propene oxidation in which cooling of the reaction gas is required to isolate the acrolein formed. Remarkably, the acrolein product of the endothermic glycerol dehydration is the same in the second step, which enhances the productivity of the process and its commercial potential.

A similar coupling has been used in the production of acrylic acid.[20] Instead of electrically cooling the reaction gas originating from the oxidation of propene to acrolein prior to its oxidation to acrylic acid, which requires a lower temperature, the oxidation of propylene to acrolein is coupled to the oxidation of acrolein to acrylic acid by including the endothermic glycerol dehydration.

3-HPA is of considerable industrial interest as an intermediate, since it is relatively easily converted to a number of large-scale commodity

Figure 5.8   At Beaumont, Texas, where Arkema is due to manufacture acrolein and
                 acrylic acid from a mixture of glycerol and propene, oil used to gush from
                 the ground in the early 1900s.
                 (Photo courtesy of the W D Hornaday Prints and Photographs Collection,
                 Texas State Library and Archives Commission.)

chemicals, including acrolein, 3-hydroxypropionic acid, acrylic acid,
malonic acid and acrylamide. 3-HPA itself exhibits antimicrobial
activity towards a wide range of pathogens and food spoilage organisms,
and it finds use both as a food preservative and as a therapeutic auxiliary
in the pharmaceutical industry. Currently 3-HPA is produced by synthesis from petrochemicals. Two chemical processes produce it as an
intermediate in the formation of 1,3-propanediol, the Degussa and Shell
processes.[21] The Degussa process starts with the catalytic transformation of propylene into acrolein, which is hydrated to 3-HPA and then
further reduced to 1,3-propanediol. The yield of 1,3-propanediol in this
process is however only 43%. The Shell process starts from ethylene and
proceeds *via* ethylene oxide, which is transformed by a hydroformyl-
ation reaction with syngas at 150 bar pressure into 3-HPA. Ethylene is
inexpensive and the intermediate products are not toxic, but the 3-HPA
has to be recovered from the organic phase. The yield of 1,3-propanediol
in this case is around 80%. Biochemical production starting from
glycerol is being intensively studied in Germany[22] and commercial
production is expected to start at Danisco Deutschland in the near
future.

References

1. *Ullmann's Encylopedia of Industrial Chemistry*, 7th edn, Weinheim: Wiley–VCH, 2006.
2. M. R. Nimlos, S. J. Blanksby, X. Qian, M. E. Himmel, D. K. Johnson, Mechanisms of glycerol dehydration. *J. Phys. Chem. A.*, 2006, **110**, 6145. Remarkably, quantum mechanical calculations in the reaction of glycerol are in agreement with the mass spectrometry observations of gas-phase fragmentation of protonated glycerol.
3. W. Girke, H. Klenk, D. Arntz, T. Haas, A. Neher, Process for the production of acrolein, US5387720 (1995).
4. C. Zhao and I. E. Wachs, Selective oxidation of propylene to acrolein over supported V_2O_5/Nb_2O_5 catalysts: An *in situ* Raman, IR, TPSR and kinetic study. *Catal. Today.*, 2006, **118**, 332.
5. J.-L. Dubois, C. Duquenne, W. Hoelderich, J. Kervennal, Process for dehydrating glycerol to acrolein, WO2006087084.
6. S. Kuba, P. Concepcion Heydorn, R. K. Grasselli, B. C. Gates, M. Che and H. Knözinger, Redox properties of tungstated zirconia catalysts : Relevance to the activation of n-alkanes. *Phys. Chem. Chem. Phys.*, 2001, **3**, 146.
7. J.-L. Dubois, C. Duquenne, W. Hoelderlich, Process for dehydrating glycerol to acrolein, WO2006087083.
8. S.-H. Chai, H.-P. Wang, Y. Liang, B.-Q. Xu, Sustainable production of acrolein: Investigation of solid acid–base catalysts for gas-phase dehydration of glycerol, *Green Chem.*, 2007, **9**, 1130.
9. E. Tsukuda, S. Sato, R. Takahashi and T. Sodesawa, Production of acrolein from glycerol over silica-supported heteropoly acids. *Catal. Commun.*, 2007, **8**, 1349.
10. I. V. Kozhevnikov, Catalysis by heteropoly acids and muticomponent in liquid-phase reactions. *Chem. Rev.*, 1998, **98**, 171.
11. S. Ramayya, A. Brittain, C. DeAlmeida, W. Mok and J. Antal, Acid-catalysed dehydration of alcohols in supercritical water. *Fuel*, 1987, **66**, 1364.
12. Conversion and acrolein selectivities which have been achieved so far are not satisfying for an economical process. L. Ott, M. Bicker, H. Vogel, Catalytic dehydration of glycerol in sub- and supercritical water: A new chemical process for acrolein production. *Green Chem.*, 2006, **8**, 214.
13. M. Watanabe, T. Iida, Y. Aizawa, T. M. Aida and H. Inomata, Acrolein synthesis from glycerol in hot-compressed water. *Bioresour. Technol.*, 2007, **98**, 1285.

14. W. Buhler, E. Dinjus, H. J. Ederer, A. Kruse and C. Mas, Ionic reactions and pyrolysis of glycerol as competing reaction pathways in near- and supercritical water. *J. Supercritical Fluids*, 2002, **22**, 37.
15. W. Hölderich, C. Duquenne, J.-L. Dubois, Method for producing acrylic acid from glycerol, WO2006114506.
16. J. E. Vancauwenberge, P. J. Slininger and R. J. Bothast, Bacterial conversion of glycerol to beta-hydroxypropionaldehyde. *Appl. Environ. Microbiol.*, 1990, **56**, 329.
17. F. Barbirato, P. Soucaille, C. Camarasa and A. Bories, Uncoupled glycerol distribution as the origin of the accumulation of 3-hydroxypropionaldehyde during the fermentation of glycerol by *Enterobacter agglomerans*. CNCM 1210. *Biotechnol. Bioeng.*, 1998, **58**, 303.
18. Q. Lüthi-Peng, S. Schärer, Z., Production and stability of 3-hydroxypropionaldehyde in *Lactobacillus reuteri*. *Appl. Microbiol. Biotechnol.*, 2002, **60**, 73.
19. J.-L. Dubois, Acrolein preparation method, WO2007090990.
20. J.-L. Dubois, Acrylic acid preparation method, WO2007090991.
21. S. Vollenweider and C. Lacroix, 3-Hydroxypropionaldehyde: applications and perspectives of biotechnological production. *Appl. Microbiol. Biotechnol.*, 2004, **64**, 16.
22. At the Institute of Bioprocess and Biosystems Engineering of the Hamburg University of Technology.

CHAPTER 6
Etherification

6.1 Etherification of Glycerol

Glycerol ethers of interest include the compounds resulting from reaction with isobutylene or *tert*-butanol, including polyglycerols and glycosyl glycerol.

6.2 Butylation to Glycerol *tert*-butyl Ethers

Glycerol cannot be added directly to fuel since its high polarity makes it is virtually insoluble in conventional fuels. In addition, its hygroscopic properties make it unsuitable as a fuel additive in unmodified form. Moreover, glycerol polymerizes at high temperatures, causing it to clog internal combustion engines and become partially oxidised into toxic acrolein. On the other hand oxygenated molecules such as methyl *tert*-butyl ether (MTBE, now banned in many US states) are valuable additives due to their antidetonant and octane-enhancing properties. In particular, glycerol *tert*-butyl ethers (GTBEs) are excellent additives and offer considerable potential for diesel and biodiesel reformulation. A mixture of 1,3-di-, 1,2-di- and 1,2,3-tri-*tert*-butyl glycerol, which is soluble in nonpolar fuels, can be incorporated into standard diesel fuel containing 30–40% aromatics. This provides significantly reduced emissions of particulate matter, hydrocarbons, carbon monoxide and unregulated aldehydes.[1] GTBEs produce a high fuel octane number, but being a branched molecule, cause a drop in cetane number.

Oxygenated diesel fuels are of importance for both environmental compliance and efficiency of diesel engines.[2] In general, the addition of these ethers has a positive effect on the quality of diesel fuel and assists in the reduction of fumes and particulates, oxides of carbon and carbonyl

RSC Green Chemistry Book Series
The Future of Glycerol: New Uses of a Versatile Raw Material
By Mario Pagliaro and Michele Rossi
© Mario Pagliaro and Michele Rossi 2008

compounds in engine exhausts. Furthermore, in the case of biodiesel a limitation in its use is the cloud point, which is $-16\,^{\circ}$C for petroleum-based diesel but around $0\,^{\circ}$C for biodiesel. The addition of ethers such as GBTEs decreases the cloud point of diesel fuels.[3] Remarkably, glycerol ethers can also considerably reduce the NO_x emissions generated from the burning of pulverized coal in heat-producing units.[4]

As already mentioned, according to an EU directive 5.75% of the total amount of fuel consumed in the Union by 2010 will have to originate from renewable sources. In Germany alone this would be equivalent to 30 million tonnes of biodiesel, and hence 3 million tonnes of glycerol, or 10 million tonnes of GTBEs. If achieved, this could be easily absorbed by the market since large quantities of isobutylene are already available due to its use as a starting material for MTBE, which although banned in California, New York and 18 other US states is still allowed in the EU.

Glycerol alkyl ethers are readily synthesized by etherification (O-alkylation) of alkenes, particularly isobutylene (Figure 6.1), in the presence of an acid catalyst at temperatures from 50 to $150\,^{\circ}$C and at molar ratios of glycerol : isobutylene of 1 : 2 or above.[5,6] The yield can be maximised by optimizing the reaction conditions, in particular the temperature, molar ratio, and type and amount of catalyst. Initially the reaction mixture consists of a two-phase system, one being a glycerol-rich polar phase

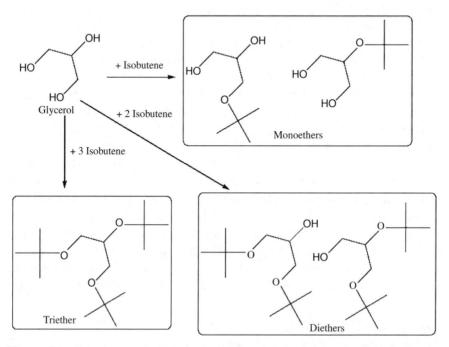

Figure 6.1 Reaction products from the *tert*-butylation of glycerol with isobutene.

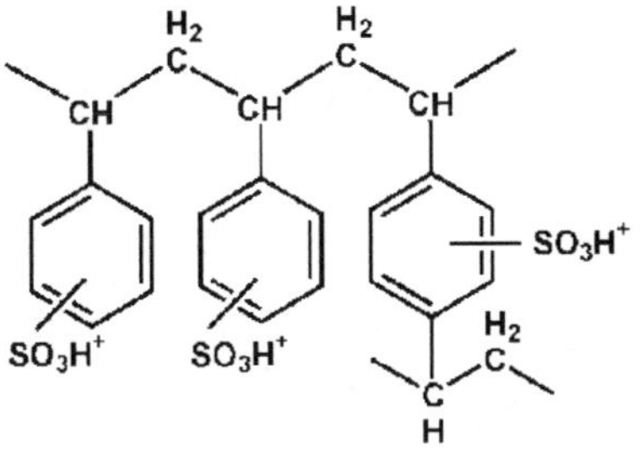

Figure 6.2 Amberlyst 35 is an excellent catalyst for glycerol alkylation with isobutene, but it requires the use of pure glycerol.

(containing the acidic catalyst and mono-*tert*-butyl glycerol ether), and the other being an olefin-rich hydrocarbon phase, also containing glycerol diethers and triether. As the reaction proceeds the products accumulate in one or other of the phases according to their solubility. During the reaction the solubility changes as the concentration of reaction products increases. At a glycerol conversion of about 60–70% the two phases amalgamate to form a single phase.

The etherification of glycerol with isobutylene in the liquid phase without solvent, catalyzed by strongly acidic resins such as Amberlyst or large-pore zeolites, affords complete conversion of glycerol.[7] Macro-reticular resins are highly active due to their large pore diameter, which remains constant due to the high level of crosslinking after swelling in the reaction medium.

Complete conversion of glycerol, with selectivity to di- and tri-ethers greater than 92%, is easily obtained over Amberlyst 35 at 60 °C (Figure 6.2). In general, five ethers are formed in the process and optimal selectivity toward ethers is achieved at 80°C with an isobutene : glycerol molar ratio around 3 : 1.[8] The mono, di- and tri-*tert*-butyl ether reaction products have been characterized using MS, NMR, IR and Raman molecular spectroscopy.[9] A drawback to the use of Amberlyst ion-exchange resins is that the methanol, salt and water present in crude glycerol originating from biodiesel production must be removed in order to avoid poisoning the catalyst.[10]

6.3 Polymerization to Polyglycerol

Polyglycerol is a highly branched polyol (Figure 6.3).

It is a clear viscous liquid, highly soluble in water and in polar organic solvents such as methanol, and is essentially nonvolatile at room temperature. At room temperature it is highly viscous, and the viscosity increases with molecular weight. Its high functionality, in combination

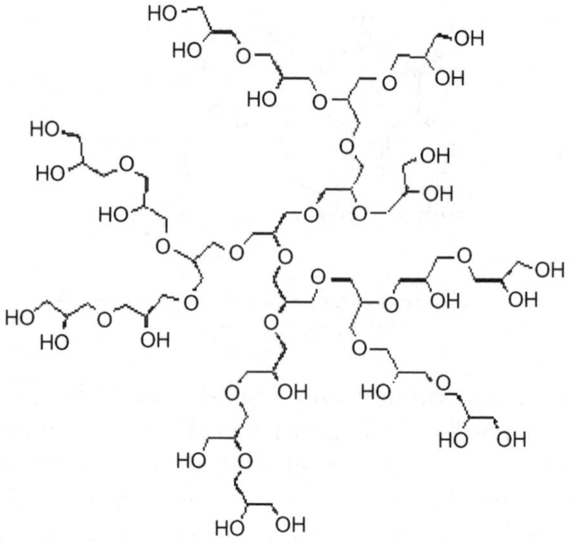

Figure 6.3 Polyglycerol is a highly branched polyol with a range of advanced applications.

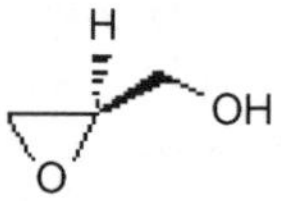

Figure 6.4 Glycidol, a glycerol derivative, is used in the controlled synthesis of polyglycerol.

with the versatile and well-investigated reactivity of hydroxyl functions, forms the basis for a variety of derivatives. A number of polyglycerols are commercially available (*e.g.*, from Hyperpolymers, Germany) for applications ranging from cosmetics to controlled drug release. Biocompatibility is an attractive feature of aliphatic polyether structures containing hydroxyl end-groups, including polyglycerols or linear polyethylene glycols (PEGs), which are approved for a wide variety of medical and biomedical applications. Controlled etherification of glycerol to form polyglycerols occurs *via* anionic polymerization of glycidol (Figure 6.4) through a rapid cation exchange equilibrium, affording polyglycerol with a narrow molecular weight distribution (polydispersity M_w/M_n is usually below 1.5) and molecular weight in the range 1000 to 30 000 g mol^{-1}.[11] Partially deprotonated (10%) 1,1,1-tris (hydroxymethyl)propane (TMP) is used as an initiator for the anionic polymerization carried out under slow addition conditions, when polymerization or cyclization is minimized in the absence of initiator.

Hyperbranched polyglycerol possesses an inert polyether scaffold. Each branch ends in a hydroxyl function, which renders hyperbranched polyglycerol a highly functional material; for example, a molecule with a molecular weight of $5000\,g\,mol^{-1}$ possesses 68 hydroxyl end-groups. As mentioned above, this high functionality, in combination with the reactivity of hydroxyl functions, forms the basis for a variety of derivatives. Partial esterification of polyglycerol with fatty acids[12] yields amphiphilic materials which behave as nanocapsules.[13] Such nanocapsules can for example incorporate polar molecules as guests and solubilize them within a nonpolar environment. Selective modification can be achieved utilizing the reactivity of 1,2-diol units, located preferentially at the periphery of the molecule.

6.4   Glycosylation to Glucosyl Glycerol

O-α-D-Glucosyl glycerol (Glc–GL) is found in Japanese traditional fermented foods such as sake, miso and mirin.[14] For example, Glc–GL contributes to the flavor of sake, which contains approximately 0.5% of this compound. Glc–GL is a non-reducing glucoside exhibiting about half the sweetness of sucrose; it has high thermal stability, low heat-discoloration, low Maillard reactivity, low hygroscopicity, high water retention capacity, high digestibility, and it is non-carcinogenic. Its use in foods and beverages is expected to increase, since it is likely to prove useful in reducing caloric intake.[15] Glc–GL is produced by an enzymatic process based on *Candida tropicalis*, an α-glucosidase with starch as donor substrate,[16] or cyclodextrin glucanotransferases,[17] which transfer the Glc residue of starch and dextrins to the 1- or 3-position of glycerol. Concentrations of 30% (w/v) glycerol and 20% (w/v) soluble starch are the most effective for ensuring efficient transglycosylation.

6.5   Industrial Applications

Glycerol ether formulations are already commercialized as oxygenate fuel additives for use in gasoline engines. For example, the US-based company CPS converts glycerol by-products originating from both the biodiesel and ethanol industries to ethers, using olefins supplied by petrochemical producers (Figure 6.5).[18] The company believes that a combination of low molecular weight olefins and glycerol, and the paraffins derived from decarboxylated fatty acids, should provide greater market penetration than either on its own.

Unlike the unpleasant tasting and water-soluble MTBE, GTBEs are insoluble in water and are only modestly toxic. For this reason

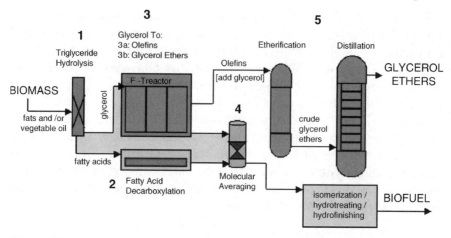

Figure 6.5 In the USA, CPS Biofuels uses glycerol produced as a by-product of biodiesel for the synthesis of glycerol ethers.

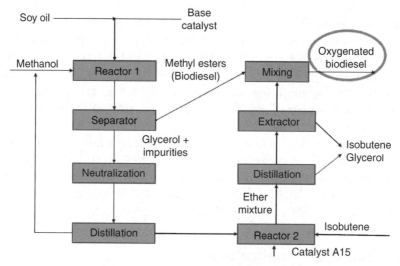

Figure 6.6 Integrated etherification will take place in the same plant as biodiesel production (Noureddini, 2001).

optimization engineering is being carried out on GTBEs both in Europe[19] and the USA.[20] In Europe, for example, the company Procede Twente is conducting what is described as a "GTBE project". The GTBE acronym for the commercial product has been trademarked and the company has joined another partner to form the GTBE Company NV in order to commercialize the technology. A fairly large demonstration plant is currently at the engineering stage, and the "GTBE: A renewable remedy for diesel soot emissions" project, funded within the European Eureka network by the Dutch and UK governments, is due to be completed by mid-2008. In general, once glycerol has been turned into a biodiesel additive the overall economics of biodiesel improve considerably since glycerol converted to GBTE is recovered in its entirety, thereby enhancing the viability of blended biodiesel.

In this context, one may easily envisage[10] a situation in which the original homogeneous manufacturing process of biodiesel could be replaced by the more efficient heterogeneous conversion, such as that of Yellowdiesel (Chapter 1). This process produces a stream of pure glycerol which can be converted to GBTE and used as a biodiesel fuel additive in one and the same plant (Figure 6.6).

References

1. F. J. Liotta, Jr., L. J. Karas, H. Kesling, Diesel fuel, US5308365 (1994).
2. F. Ancillotti and V. Fattore, Oxygenate fuels: Market expansion and catalytic aspect of synthesis *Fuel Proc. Tech.*, 1998, **57**, 163.
3. H. Noureddini, Process for producing biodiesel fuel with reduced viscosity and a cloud point below 32°F, US6174501 (2001).
4. D. B. Appleby, J. K. Spooner-Wyman, Method for combustion of pulverized coal with reduced emissions, US7195656 (2007).
5. V. P. Gupta, Glycerine ditertiary butyl ether preparation, US5476971 (1995).
6. A. Behr and L. Obendorf, Development of a process for the acid-catalyzed etherification of glycerine and isobutene forming glycerine tertiary butyl ethers. *Engin. Life Sci.*, 2002, **2**, 185.
7. (a) K. Klepacova, D. Mravec and M. Bajus, *tert*-Butylation of glycerol catalysed by ion-exchange resins *Appl. Catal. A.*, 2005, **294**, 141; (b) K. Klepacova, D. Mravec and M. Bajus, Etherification of glycerol with *tert*-butyl alcohol catalysed by ion-exchange resins. *Chem. Papers*, 2006, **60**, 224.
8. R. S. Karinen and A. O. I. Krause, New biocomponents from glycerol. *Appl. Catal. A.*, 2006, **306**, 128.
9. M. E. Jamróz, M. Jarosz, J. Witowska-Jarosz, E. Bednarek, W. Tęcza, M. H. Jamróz, J. Cz. Dobrowolski and J. Kijeński, Mono-, di-, and tri-*tert*-butyl ethers of glycerol: A molecular spectroscopic study. *Spectrochim. Acta A-Mol. Biomol. Spectrosc.*, 2007, **67**, 980.
10. H. Noureddini, W. R. Dailey, B. A. Hunt, Production of ethers of glycerol from crude glycerol, 1998, http://digitalcommons.unl.edu/chemeng_biomaterials/18.

11. A. Sunder, R. Hanselmann, H. Frey and R. Mülhaupt, Controlled synthesis of hyperbranched polyglycerols by ring-opening multi-branching polymerization. *Macromolecules*, 1999, **32**, 4240.
12. Elegant shape-selective catalysts enable control of crucial degree of esterification. See C. Márquez-Alvarez, E. Sastre, J. Pérez-Pariente, Solid catalysts for the synthesis of fatty esters of glycerol, poly-glycerols and sorbitol from renewable resources. *Top. Catal.* 2004, **27**, 105.
13. A. Sunder, M. Krämer, R. Hanselmann, R. Mülhaupt and H. Frey, Molecular nanocapsules based on amphiphilic hyperbranched polyglycerols. *Angew. Chem. Int. Ed.*, 1999, **38**, 3552.
14. F. Takenaka and H. Uchiyama, Synthesis of α-D-glucosylglycerol by α-glucosidase and some of its characteristics. *Biosci. Biotechnol. Biochem.*, 2000, **64**, 1821.
15. F. Takenaka and H. Uchiyama, Effect of α-D-glucosylglycerol on the *in vitro* digestion of disaccharides by rat intestinal enzymes. *Biosci. Biotechnol. Biochem.*, 2001, **65**, 1458.
16. T. Sawai and E. J. Hehre, A novel amylase (*Candida transglucosyl-amylase*) that catalyzes glucosyl transfer from starch and dextrins. *J. Biol. Chem.*, 1962, **237**, 2047.
17. H. Nakano, T. Kiso, K. Okamoto, T. Tomita, M. B. A. Manan and S. Kitahata, Synthesis of glycosyl glycerol by cyclodextrin glucano-transferases. *J. Biosci. Bioeng.*, 2003, **95**, 583.
18. By the CPS Biofuels company (Chapel Hill, NC): http://www.cpsbiofuels.com.
19. By the Netherlands-based consortium Procede Group BV, with the involvement of industrial partners.
20. J. K. Spooner-Wyman, D. B. Appleby, D. M. Yost (2003) Evaluation of di-butoxyglycerol (DBG) for use as a diesel fuel blend component, *SAE Technical Paper Series*, Paper No. 2003-01-2281. Despite the paper's title, the industrial chemist authors of this paper deal with glycerol ethers produced by the reaction of isobutylene and glycerol, namely di-*tert*-butyl ethers.

CHAPTER 7
Esterification

7.1 Esterification of Glycerol

Esterification of glycerol affords a variety of useful products and in recent years has been an active area of research. Reactions employ both chemical catalysts and lipases and can be divided into three types: esterification with carboxylic acids, carboxylation, and nitration.

7.2 Esterification with Carboxylic Acids and Glycerolysis

Esterification of glycerol with carboxylic acids results in mono-acylglycerols (MAGs) and diacylglycerol (DAG) (Figure 7.1). MAGs are amphiphilic molecules and are useful as nonionic surfactants and emulsifiers.

Both MAGs and DAGs are widely used as food additives in dairy and bakery products, margarines and sauces. They assist in blending certain ingredients together, such as oil and water, which would otherwise be difficult to blend.[1] In the cosmetics industry they are employed as texturing agents for improving the consistency of creams and lotions.[2] In addition, owing to their excellent lubricant and plasticizing properties, MAGs are used in textile processing oils for application on various types of machinery.[3] They are currently manufactured industrially either by continuous chemical glycerolysis of fats and oils at high temperature (220–250 °C), employing alkaline catalysts under a nitrogen atmosphere, or by the direct esterification of glycerol with fatty acids.[4] These catalytic processes require strongly basic catalysts such as KOH, NaOH or Ca(OH)$_2$ and lead to the formation of monoglycerides of limited purity, due to the presence of by-products such as di- and triglycerides and

RSC Green Chemistry Book Series
The Future of Glycerol: New Uses of a Versatile Raw Material
By Mario Pagliaro and Michele Rossi
© Mario Pagliaro and Michele Rossi 2008

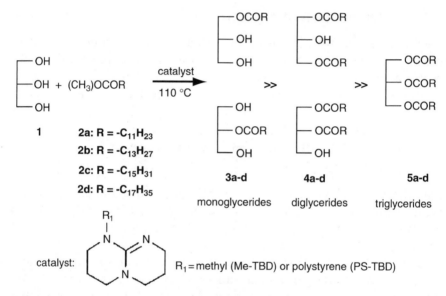

Figure 7.1 Monoacylglycerol and diacylglycerol structures.

Figure 7.2 Catalytic esterification of unprotected glycerol over guanidine derivatives. (Reproduced from Ref. 6, with permission.)

soaps. For this reason MAGs destined for food use must undergo a distillation step.[5] An alternative approach uses a supported guanidine catalyst to provide a one-pot glycerol ester synthesis with high yield and high monoglyceride selectivity, starting from a glycerol : methyl ester molar ratio of 1 : 1 (Figure 7.2).[6]

In Figure 7.2, starting from **2a,** after 50 h reaction glycerol esters were produced in more than 98% yield with a very high selectivity (78%), to give **3a**. Alternative methods involve immobilized lipases. An example is the synthesis of MAGs using *Staphylococcus simulans* lipase immobilized on $CaCO_3$ in a solvent-free system,[7] giving a 70.6% yield of pure MAGs suitable for food use.

Catalytic glycerolysis of a triglyceride yields mono- and diglycerides, eqns (1) and (2) respectively:

$$\text{triglyceride} + 2\,\text{glycerol} \longleftrightarrow 3\,\text{monoglyceride} \qquad (1)$$

$$2\,\text{triglyceride} + \text{glycerol} \longleftrightarrow 3\,\text{diglyceride} \qquad (2)$$

Olive oil and glycerol, for example, react over immobilized Novozyme lipases at 80 °C and atmospheric pressure to give good yields of glycerol dioleate (Figure 7.3).[3]

However, an excess of glycerol does not result in a significant increase in monoglyceride yield, and similar product distributions are obtained for glycerol : oil mole ratios of 3 : 1 and 6 : 1.

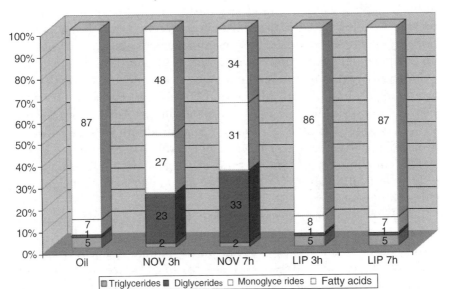

CATALYST	t (h)	FFA	MG	DG	TG
Olive oil	0	5	1	7	87
Novozyme 435	3	2	23	27	48
Novozyme 435	7	2	33	31	34
Lipozyme IM	3	5	1	8	86
Lipozyme IM	7	5	1	7	87

Figure 7.3 Comparison between different lipases for the transesterification reaction between olive oil and glycerol. (Reproduced from Ref. 3, with permission.)

Table 7.1 Comparison between lipase- and base-catalyzed transesterification of olive oil with glycerol. (Reproduced from Ref. 8, with permission.)

PROCESSES	REACTION CONDITIONS	DOWNSTREAM PROCESSES	REFINED PROCESSES
TRANSESTERIFICATION BASIC CATALYSIS (Glycerol monooleate)	180 °C Atmospheric pressure 0.2% wt catalyst	1. DECANTATION 2. WASHING	1. SHORT PATH DISTILLATION
TRANSESTERIFICATION ENZYMATIC CATALYSIS (Glycerol monooleate)	60 °C Atmospheric pressure 1.5% wt catalyst	CRYSTALLYZATION (3 STEPS)	1. DEODORIZATION 2. DRYING

Comparison of the traditional and enzymatic processes for the production of MAG clearly shows the key advantages of the bio-process in terms of higher selectivity (and less waste) and lower temperature and pressure (Table 7.1). Since the enzyme costs can be reduced by a factor of three by economies of scale, this and similar lipase-based processes will find commercial application in large-scale transesterification reactions.

Generally speaking, the methodology is versatile and can be applied to the preparation of both MAGs and DAGs starting from fatty acids. For example, esterification of glycerol with linoleic acid in hexane over 15% (w/w) lipase from *Rhizomucor miehei* produces 80% esterification to 1,3-diacylglycerol and 1-monoacylglycerol after 8 h at 50 °C in a system containing a 1 : 2 molar ratio of glycerol : free fatty acids; similarly, esterification levels > 80% are obtained in 8 h at 40 °C using lipase from *Candida antarctica*.[9]

DAG is present naturally as a minor component of edible fats and oils from various sources.[10] It can exist as either 1,3-DAG or 1,2 (2,3)-DAG. DAG has been utilized as a cocoa butter blooming agent and as an intermediate for the synthesis of structural lipids.[11] Recent studies of its nutritional properties and dietary effects suggest that DAG in which 1,3-DAG is the major component can play a role in reducing serum triacylglycerol (TAG) levels, and as a result is able to decrease body weight and visceral fat mass.[12] Consequently, oils with a high DAG content have gained attention for the prevention of obesity and other lifestyle-related conditions.[13] State of the art production of 1,3-DAG using lipases is based on esterification of glycerol with fatty acids using a solvent-free approach (Figure 7.4).[14] This is an advance on traditional esterification in organic solvent.[15]

On the laboratory scale an 84% yield of DAG with a purity of 90% is obtained using 1.29 mM glycerol and 2.59 mM fatty acid using a

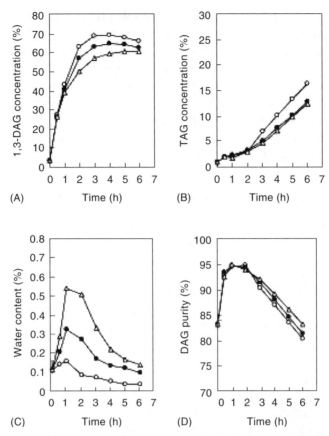

Figure 7.4 Effect of vacuum conditions on the production of 1,3-DAG: (A) 1,3-DAG;
(B) TAG; (C) water content in the reaction mixture; and (D) DAG purity.
The reaction was performed at 50 °C with a molar ratio of fatty acids to
glycerol of 2.0 and a residence time of 60 s. The Lipozyme RM IM concentration was 5% (dry weight basis): (O) 0.4 kPa; (●) 0.8 kPa; and (Δ)
1.3 kPa. (Reproduced from Ref. 14, with permission.)

1,3-regioselective lipase in a solvent-free system. Due to the low temperature of the reaction (25 °C), non-solvent esterification requires a
relatively long time (*e.g.* 12 h) to produce dilinolein (71.7%). A better
approach for production on the industrial scale is to use a temperature
of 50 °C with continuous removal of water.[16] In particular, a packed bed
reactor containing an immobilized 1,3-regioselective lipase, connected to
a water removal vessel under optimized vacuum conditions of 0.4 kPa
and 50 °C, yields a maximum 1,3-DAG content around 70%, provided
the molar ratio of fatty acid to glycerol is above 2.0.[17]

Under 0.4 kPa the DAG purity reaches a maximum of 95.4% after
2 h reaction, and then gradually decreases to 81.2% after 6 h (D in

Figure 7.4) due to increasing TAG concentration in the later stages of the reaction. Although the rate of production of 1,3-DAG appears to be constant regardless of the vacuum conditions at the initial stages of the reaction, its concentration in the later stages increases under higher vacuum (A). Water is thus removed more effectively under higher vacuum conditions. For example, the water content after 6 h reaction is 0.057% at 0.4 kPa (C). The reason for the lower rate of 1,3-DAG synthesis at later stages of the reaction under reduced vacuum is due to an increase in the reverse reaction (hydrolysis).

7.3 Carboxylation to Glycerol Carbonate

Glycerol carbonate (4-hydroxymethyl-1,3-dioxolan-2-one; GC) is a colorless protic polar liquid. Being non-toxic and having a high boiling point, GC is a useful solvent for plastics and resins such as cellulose acetate, nylon, nitrocellulose and polyacrylonitrile. It also has potential as bio-lubricant, due to its adhesion to metallic surfaces and resistance to oxidation, hydrolysis and pressure. It can be prepared directly and in high yield from glycerol and dimethyl carbonate in a reaction catalyzed by lipases (Figure 7.5).

Another method of obtaining carbonate derivatives of glycerol is by reaction between urea and glycerol using a mineral zinc sulfate catalyst, producing GC with good productivity and selectivity (92%) by carbamoylation–carbonation at a temperature between 90 and 220 °C in the presence of a Lewis acid catalyst. The second, slower, step requires a salt such as zinc sulfate as catalyst (Figure 7.6).

The reaction is carried out at a pressure of 40–50 mbar in order to shift the equilibrium by eliminating ammonia in the gaseous phase. Synthesis and purification of glycerol carbonate in a thin-film separation 2 L thermal reactor has produced yields of 94% and 97%.[19] Other methods involve transesterification with ethylene carbonate or dialkyl carbonates;[20] for example, in a reaction with ethylene carbonate carried

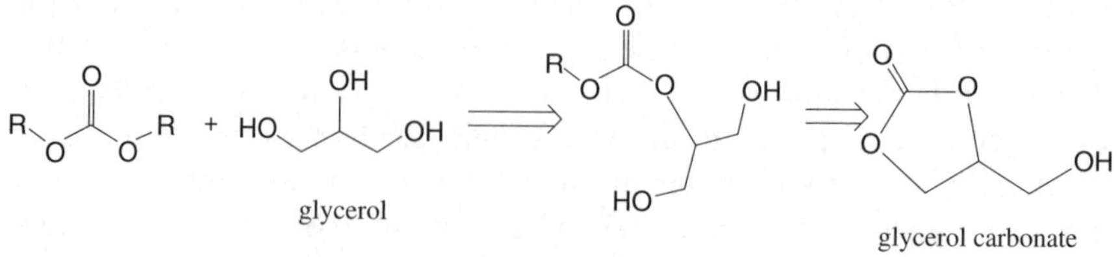

glycerol

glycerol carbonate

Figure 7.5 GC can be produced from the reaction of dialkylcarbonate (R = alkyl) with glycerol, with the intermediate ester undergoing a second esterification.

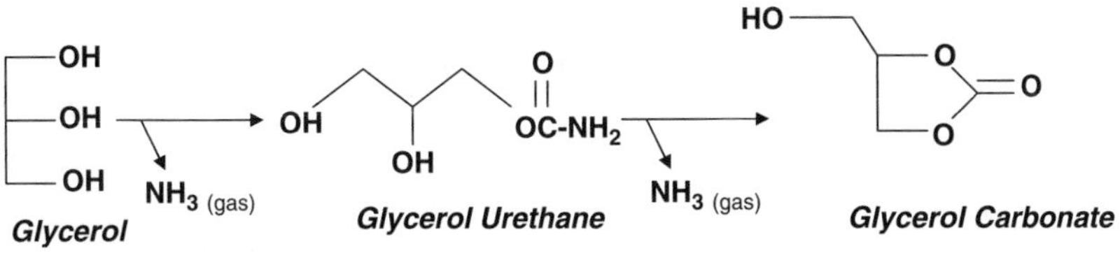

Figure 7.6 Reaction of urea and glycerol catalyzed by Zn^{2+} at reduced pressure yields GC. (Reproduced from Ref. 18, with permission.)

out at 125 °C in the presence of sodium bicarbonate, GC was formed in 81% yield.

When heated GC reacts readily with phenols, alcohols and carboxylic acids to form ethers or esters of glycerol, including polymers such as polyesters, polycarbonates, polyurethanes and polyamides. Being relatively inexpensive, GC could serve as a source of new polymeric materials based on glycidol, which is readily obtained in high yield from GC by catalytic reaction within the pores of zeolite A. Reaction at 180 °C and at pressures as low as 35 mbar produces glycidol in 86% yield and 99% purity, by contraction of the five-membered cyclic carbonate unit into a three-membered cyclic epoxy unit (Figure 7.7).[21]

Glycidol is a clear slightly viscous liquid with a variety of industrial uses. It is used in surface coatings and as a gelation agent in solid propellants, as a stabilizer for natural oils and vinyl polymers, and as a demulsifier. It is used as a chemical intermediate in the synthesis of glycidyl ethers, esters and amines and as a high-value component in the production of epoxy resins and polyurethanes. For example, glycidyl carbamates, synthesized by the reaction of polyfunctional isocyanate oligomers and glycidol, combine the excellent properties of polyurethanes with the crosslinking chemistry of epoxy resins.[22] Glycidol is itself polymerized commercially to polyglycerol (Chapter 6).[23]

7.4 Nitration

The explosive nitroglycerine, currently produced in modular plants, is also successfully employed as an antianginal drug (Figure 7.8).

When treated with nitrating agents glycerol forms a solution containing dinitroglycerol. If the solution is then treated with a cyclizing agent the dinitroglycerol is converted into glycidyl nitrate (Figure 7.9), which can be polymerized to poly(glycidyl nitrate) (PGN).

The polymer PGN is potentially suitable for use in propellants, explosives, gas generators and pyrotechnics. Its industrial synthesis typically follows a three-step procedure, involving nitration of epichlorohydrin

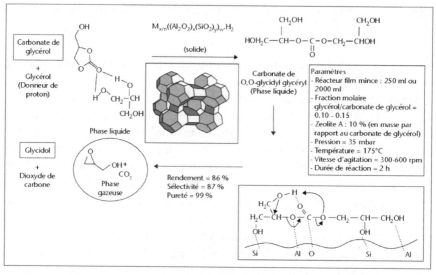

Figure 7.7 Glycidol is readily formed by oligomerisation of GC, catalyzed under reduced pressure over zeolite A or γ-alumina. (Reproduced from Ref. 18, with permission.)

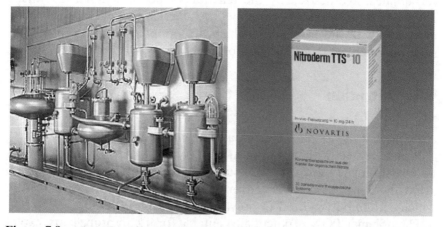

Figure 7.8 Nitroglycerine, currently produced in modular plants (*left*) is also successfully employed as an antiangina drug (*right*). (Reproduced from Novartis.com, with permission.)

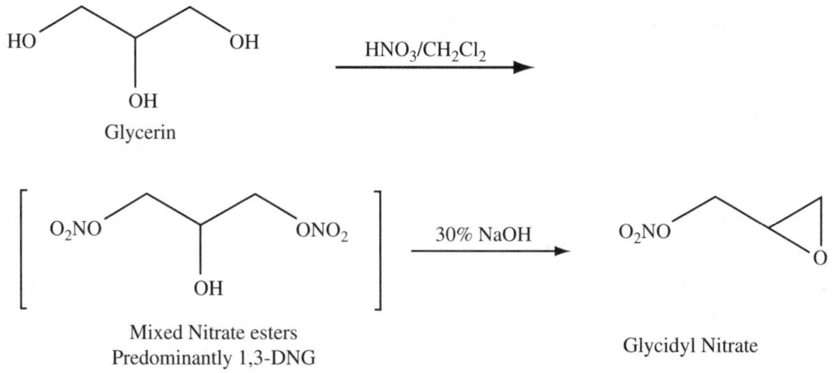

Figure 7.9 Continuous process for the production of glycidol nitrate, and thus PGN, from glycerol involves consecutive nitration and causticication.

(now obtained from glycerol), followed by cyclization of the nitrated epichlorohydrin using a base, to form glycidyl nitrate. In the third step the glycidyl nitrate is polymerized cationically to form PGN. This method of producing glycidyl nitrate is however dangerous and is not economically feasible for large-scale commercial production. A safe and relatively inexpensive process for producing glycidyl nitrate in large quantities, at sufficient purity for conversion to PGN without further distillation, has therefore been developed.[24] Glycerol and nitric acid are continuously reacted in a first reaction vessel to produce 1,3-dinitroglycerol in high yield (at least 50%), assisted by the asymmetrical kinetics of the continuous process. The outflow is reacted with excess NaOH in a second reaction vessel to neutralize the excess nitric acid and to yield glycidyl nitrate. The outflow from this vessel passes into a decanter, which separates the immiscible organic phase from the basic aqueous solution. The temperature of the nitration reaction is around 5 °C and the preferred temperature for causticization and separation is about 25 °C. This offers considerable advantages over conventional glycidyl nitrate processes, typically carried out at temperatures as low as −10 to −70 °C. The addition of an organic solvent such as dichloromethane dilutes and moderates the reactions, and provides a significant measure of safety to the process by absorbing the heat of reaction, since it boils before a dangerous temperature is reached. In addition, the boiling organic solvent drives off the catalyst decomposition product, oxides of nitrogen (NO_x), stabilizing the reaction and making it suitable for commercial use. The glycidyl nitrate produced is pure enough for direct use in the production of PGN. This avoids the hazardous and costly purification required by other glycidyl nitrate processes.

7.5 Industrial Applications

Glycerol esters such as MAGs and DAGs have well established industrial applications, particularly in the food and oleochemical industries. In 2006 the world market for food emulsifiers was estimated to be around 400 000 tons, approximately 10% of which was monoolein. Of the 40 emulsifier units across the world, 25 are located in the United States. DAG has the advantage of being stable to decomposition at cooking temperatures. It is produced by the Japanese company Kao from soybean and canola oil using a lipase[16] and marketed under the tradename "Healthy Econa Oil" for use in cooking, frying and dressings (Figure 7.10).[25]

DAG is in fact present at a level of a few percent in virtually all edible oils, whether vegetable or animal, and has been long ingested by man.[10] Research concerning the human nutritional characteristics of DAG oil compared to triacylglycerol (TAG) oil clearly demonstrates the significant suppressive effect by DAG on body fat accumulation.[26] This is due to the reduced possibility of synthesis of TAG in the small intestine following DAG oil digestion. This led Kao to start production of DAG-based oil, aimed at an affluent Japanese population of 128 million for which obesity has become a serious issue. Since its introduction in 1999 Econa has become the best selling vegetable oil in the Japanese market, due to its health benefits and its mild flavor. In a joint venture with Kao, Archer Daniels Midland manufactures and markets Econa in the Americas, Europe, Australia and New Zealand.

The new process for production of glycidyl nitrate is already in use in the USA by the weaponry manufacturer Alliant Techsystems. The continuous nature of the process reduces labor costs, and the conventional equipment used, its inherent safety, increased nitration yields, lack of

Figure 7.10   1,3-DAG content of Kao Econa cooking oil is 80%. (Reproduced from Kao.com, with permission.)

hazardous waste streams, thermal savings and the improved purity of the product, all contribute to its technical and economic viability. Finally, in France the company Condat is about to commercialize products in the form of glycerol polycarbonates and glycerol carbonate polyesters, obtained as the result of three one-pot reactions: carbonation, oligo-merization and acylation. This opens the way to new multifunctional plant-based polymers, which have already proved themselves effective as industrial lubricants and biodegradable hydraulic fluids that are both non-toxic and fire-resistant. With their anti-wear properties, even at high pressures, as well as their anti-friction characteristics, they will find application in metal working and machining.

References

1. M. A. Jackson and J. W. King, Lipase-catalyzed glycerolysis of soybean oil in supercritical carbon dioxide. *J. Am. Oil Chem. Soc.*, 1997, **74**, 103.
2. D. E. Stevenson, R. A. Stanley and R. H. Furneaux, Glycerolysis of tallow with immobilised lipase. *Biotechnol. Lett.*, 1993, **15**, 1043.
3. M. Martinez, A. Coterón and J. Aracil, Reactions of olive oil and glycerol over immobilized lipases. *J. Am. Oil Chem. Soc.*, 1998, **75**, 657.
4. N. O. V. Sonntag, Glycerolysis of fats and methyl esters: Status, review and critique. *J. Am. Oil. Chem. Soc.*, 1992, **59**, 795.
5. U. T. Bornscheuer and T. Yamane, Fatty acid vinyl esters as acylating agents: A new method for the enzymatic synthesis of monoacylglycerols. *J. Am. Oil Chem. Soc.*, 1995, **72**, 193.
6. F. Jérôme, G. Kharchafi, I. Adam and J. Barrault, One pot and selective synthesis of monoglycerides over homogeneous and heterogeneous guanidine catalysts. *Green Chem.*, 2004, **6**, 72.
7. H. Ghamgui, N. Miled, A. Rebai, M. Karra-Chaabouni and Y. Gargouri, Production of monoolein by immobilized *Staphylo-coccus simulans* lipase in a solvent-free system: Optimization by response surface methodology. *Enzyme Microb. Technol.*, 2006, **39**, 717.
8. M. Vicente, J. Aracil and M. Martinez, *Biocatalytic processes for the production of fatty acid esters*, BREW Symposium: Bioper-spectives 2005, 11 May, 2005.
9. C. E. Martinez, J. C. Vinay, R. Brieva, C. G. Hill and H. S. Garcia, Preparation of mono- and diacylglycerols by enzymatic esterifica-tion of glycerol with conjugated linoleic acid in hexane. *Appl. Biochem. Biotechnol.*, 2005, **125**, 63.

10. R. P. D'Alonzo, W. J. Kozarek and R. L. Wade, Fats and oils as determined by glass capillary gas chromatography. *J. Am. Oil Chem. Soc.*, 1982, **59**, 292.

11. R. Haftendorn and R. Ulbrich-Hofman, Synthesis of 2-modified 1,3-diacylglycerols. *Tetrahedron*, 1995, **51**, 1177.

12. J. Nagata and M. Saito, Effects of simultaneous intakes of indigestible dextrin and diacylglycerol on lipid profiles in rats fed cholesterol diets. *Nutrition*, 2006, **22**, 395.

13. B. D. Flickinger and N. Matsuo, Nutritional characteristics of DAG oil. *Lipids*, 2003, **38**, 129.

14. R. Rosu, M. Yasui, Y. Iwasaki and T. Yamane, Enzymatic synthesis of symmetrical 1,3-diacylglycerols by direct esterification of glycerol in solvent-free system. *J. Am. Oil Chem. Soc.*, 1999, **76**, 839.

15. M. Berger and M. P. Schneider, Enzymatic esterification of glycerol: II Lipase-catalyzed synthesis of regioisomerically pure 1(3)-*rac*-monoacylglycerols. *J. Am. Oil Chem. Soc.*, 1992, **69**, 961.

16. T. Watanabe, M. Shimizu, M. Sugiura, M. Sato, J. Kohori, N. Yamada and K. Nakanishi, Optimization of reaction conditions for the production of DAG using immobilized 1,3-regiospecific lipase lipozyme RM IM. *J. Am. Oil Chem. Soc.*, 2003, **80**, 1201.

17. T. Watanabe, M. Sugiura, M. Sato, N. Yamada and K. Nakanishi, Diacylglycerol production in a packed bed bioreactor. *Process Biochem.*, 2005, **40**, 637.

18. Z. Mouloungui, Voies inhabituelles de synthèse de composés oléophiles à partir des substrats végétaux solides (graines oléoprotéagineuses), liquides (huiles végétales et dérivés, glycérol) pour l'industrie chimique *OCL*, 2004, **11**(6), 425.

19. S. Claude, Z. Mouloungui, J.-W. Yoo and A. Gaset, Method for preparing glycerol carbonate, US6025504 (1999).

20. J. B. Bell, L. Silver and V. Arthur, Method for preparing glycerin carbonate, US2915529 (1959).

21. Small scale production of these highly branched glycidol polymers has been commercialized in Europe by Hyperpolymers (Germany).

22. P. A. Edwards, G. Striemer and D. C. Webster, Synthesis, characterization and self-crosslinking of glycidyl carbamate functional resins. *Progr. Org. Coatings*, 2006, **57**, 128.

23. G. Rokicki, P. Rakoczy, P. Parzuchowski and M. Sobiecki, Hyperbranched aliphatic polyethers obtained from environmentally benign monomer: glycerol carbonate. *Green Chem.*, 2005, **7**, 529.

24. T. K. Highsmith and H. E. Johnston, Continuous process and system for production of glycidyl nitrate from glycerin, nitric acid and caustic and conversion of glycidyl nitrate to poly(glycidyl nitrate), US6870061 (2005).

25. Econa Cooking Oil is a trademark of Kao Corporation. Detailed information on this 1,3-DAG-based product is available at the URL: http://www.kao.co.jp/rd/dag_en/index.html.

26. T. Nagao, H. Watanabe, N. Goto, K. Onizawa, H. Taguchi, N. Matsuo, T. Yasukawa, R. Tsushima, H. Shimasaki and H. Itakura, Dietary diacylglycerol suppresses accumulation of body fat compared to triacylglycerol in men in a double-blind controlled trial. *J. Nutr.*, 2000, **130**, 792.

CHAPTER 8
Selective Oxidation

8.1 Selective Oxidation of Glycerol

Selective oxidation of glycerol is of particular interest on account of the commercial relevance of oxygenated glycerol derivatives. During the last decade the whole arsenal of chemical, electrochemical and biological oxidation methods has been explored for creating a market outlet for the large surplus of biodiesel glycerol. As the result of this effort, human ingenuity has now arrived at the series of glycerol derivatives shown in Figure 8.1. Previously this versatility was entirely absent, with biological transformation producing only a single product, dihydroxyacetone, which has an annual global market of no more than 2000 tonnes.

As an example of recent progress, newly developed gold catalysts now provide the basis for a highly selective process for the human metabolite, glyceric acid (Figure 8.2).

This interesting bifunctional molecule has not yet been fully developed as a chemical intermediate owing to the lack of a large-scale preparative method. On the other hand novel food supplements or medicinal products containing D-glyceric acid are now commercialized for improving ethanol metabolism.[1] In addition, conventional platinum and palladium catalysts, in newly developed bi- and polymetallic systems, are useful tools for intensive oxidation of glycerol to tartronic and mesoxalic acids. As a further example, the electrochemical oxidation of glycerol in fuel cells will without doubt provide an important large-scale use for glycerol, completing the cycle of conversion of renewable seed oil to energy. The use of glycerol in biofuel cells has considerable advantages over ethanol and methanol, which have been the traditional first choice for biofuel cells for portable devices (Figure 8.3).[2]

Progress in oxidation technology marks the irreversible decline of so-called stoichiometric oxidants such as chromates, permanganates and

RSC Green Chemistry Book Series
The Future of Glycerol: New Uses of a Versatile Raw Material
By Mario Pagliaro and Michele Rossi

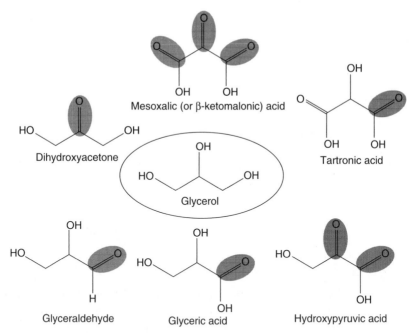

Figure 8.1 Oxidized glycerol derivatives.

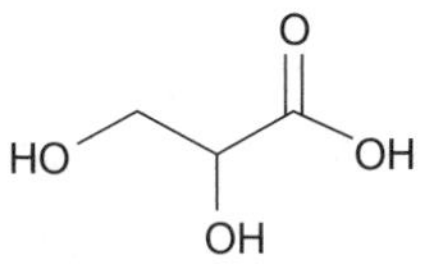

Figure 8.2 Glyceric acid is a synthon of largely unexplored potential, whose D-enantiomer is an anticirrhosis agent.

hypochlorites, which produce unacceptable types of waste, in favor of clean processes carried out under mild conditions using molecular oxygen in aqueous solution. An example of this challenging approach is the use of heterogeneous catalysis in which a solid catalyst is stirred with liquid reagents in the presence of oxygen or, even better, air. This avoids the large scale use of soluble catalysts, it simplifies separation of the catalyst from the products and, finally, it speeds purification.

The aerobic oxidation of alcoholic hydroxyl groups was developed in the early 1990s with the use of conventional Pt- and Pd-supported catalysts and has since evolved into sophisticated polymetallic systems and organocatalysis, which increase selectivity and limit catalyst deactivation.[3] Owing to the presence of three oxdizible hydroxyl groups,

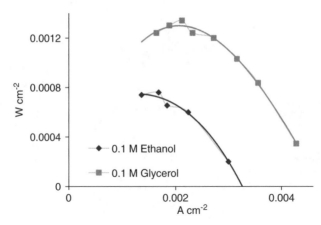

Figure 8.3 Comparison of the power curves for a single biofuel cell with two different analyte fuels (100 mM ethanol and 100 mM glycerol) at room temperature. (Adapted from Ref. 2, with permission.)

a variety of glycerol derivatives are made available by adding oxygen and removing hydrogen atoms. A selection of stable, commercially relevant oxygenated products is indicated in Figure 8.1. Thanks to the current low price of glycerol, the economy of each product is governed to a large extent by the processing costs, in which the selectivity–conversion balance plays a crucial role. While simple oxygenated derivatives of glycerol such as peroxides are practically unknown, oxidative dehydrogenation is the common route to carbonyl derivatives, which may then undergo further oxygen addition to carboxylic derivatives. Glyceraldehyde (glycerose) is of scientific interest in regard to its choice as standard for relative enantiomeric configurations and to the role of its 3-phosphate derivative as an intermediate in carbohydrate metabolism. In addition, this simple aldose undergoes the Lobry de Bruyn–Alberda van Ekenstein rearrangement, producing a mixture of glyceraldehyde and dihydroxyacetone in the biogenesis of animal organisms.[4] Owing to its high reactivity with molecular oxygen, however, glyceraldehyde itself is rarely observed among the products of the aerobic oxidation of glycerol. Dihydroxyacetone, a highly valued chemical used in artificial tanning and as a synthon for organic synthesis, is at present produced by fermentation of glycerol in the presence of *Gluconobacter oxidans*.[5] Glyceric and β-hydroxypyruvic acids are promising monomers for new polymeric materials and chelating agents, but their use in the food industry[6] is at present limited by the lack of a low-cost synthesis. Tartronic acid is available on the small scale *via* malonic acid ozonization,[7] whereas the naturally occurring mesoxalic acid, an interesting pharmaceutical precursor,[8] can be prepared by either catalytic or electrochemical methods.[9] The last two products are the precursors of interesting polyketomalonate materials.[10]

8.2 Thermodynamic and Kinetic Aspects of Aerobic C–OH Oxidation

From the thermodynamic point of view dioxygen can be considered a high-energy molecule which reacts exothermically with organic compounds. However, activation of dioxygen is of fundamental importance in selective oxidation, in fact most organic molecules are stable in the presence of dioxygen owing to the high activation energy. The stability of organic molecules towards dioxygen is a kinetic property. Molecular oxygen is characterized by its paramagnetic and electrophilic character, which can be interpreted by the molecular orbital of its fundamental state (Figure 8.4).

As a radical it can react easily with other paramagnetic species, and as an electrophile it forms charge transfer complexes which can be reduced

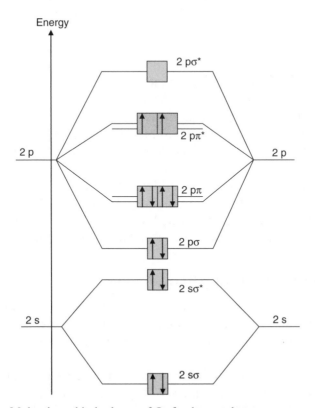

Figure 8.4   Molecular orbital scheme of O_2 fundamental state.

Figure 8.5   *gem*-Diol derived from water addition undergoes ready dehydrogenation.

by accepting up to four electrons in the high-energy orbitals. To over-come kinetic inertness two mechanisms are available. The first is the promotion of the organic substrate to the upper triplet state, and the second consists of the so-called dioxygen activation. The latter can be achieved either by light-induced promotion to an O_2 singlet state, or more commonly by interaction with a metallic atom or ion. The $M–O_2$ bond produces sufficient energy to overcome the kinetic barrier of spin coupling, whereas the spin orbit coupling energy in the metal–dioxygen complex reduces the kinetic barrier of the spin change.

Dioxygen and the hydroxylated molecule are activated on a solid catalyst surface, producing a dehydrogenation process which leads to carbonylic species. The process requires temperatures around 60–80 °C, and the carbonyl intermediates behave either as kinetically inert species or they can be further oxidized, depending on experimental conditions. While in the absence of water the oxygen attack on the aldehydic carbon is a slow process, the *gem*-diol derived from water addition readily undergoes dehydrogenation (Figure 8.5).

Hydration of the intermediate aldehyde is crucial for generating a carboxylated species. Thus, the selectivity can be towards carboxylates in aqueous solution with alkali as hydration catalyst, whereas in organic media or solvent-free conditions carbonylic species are the main products.

8.3   Platinum Group Metal Catalysis

In general, platinum and palladium are highly active aerobic catalysts for the oxidation of glycerol. High yields of dihydroxyacetone can be obtained by the aerobic oxidation of glycerol on bismuth-promoted platinum in acidic media (Figure 8.6).[11]

Oxidation at the primary carbon to produce glyceric acid is favored on platinum or palladium catalysts. Reasonably good selectivity (80%) to glyceric acid was obtained at 100% conversion with a Pd/C catalyst at

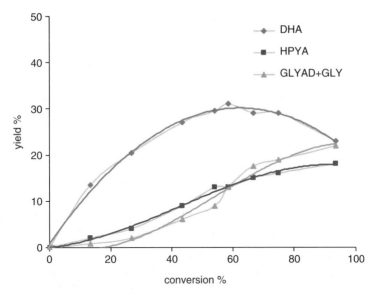

Figure 8.6 Oxidation of glycerol over Pt–Bi/C catalyst at pH 2 yields mainly DHA (GLYA = glyceric acid, GLYAD = glyceraldehyde). (Adapted from Ref. 11, with permission.).

pH 11, whereas a Pt/C catalyst produced a lower yield (50%). By prolonging the oxidation, glyceric acid could be further oxidized to tartronic acid with a reaction rate increasing with pH, the optimal yield being observed at pH $\geq$ 9.

In addition, tartronic acid can be further oxidized to mesoxalic acid with air in acidic media on a Pt–Bi/C catalyst, but the yield of this step is modest (60% yield at 80% conversion). To avoid the complex separation of mesoxalic acid from the tartronic acid it may be advantageous to sacrifice some yield in order to drive the conversion to 100%. In this case a 50% yield of mesoxalic acid, contaminated by traces of oxalic acid, is finally obtained.[12] Bismuth-promoted platinum, in its turn, mainly catalyzes the conversion of glyceric acid at low pH to β-hydroxypyruvic acid (64% yield at 75% conversion).[13]

While a Bi–Pt catalyst shows high selectivity to dihydroxyacetone, a three-component Ce–Bi–Pd catalyst supported on carbon allows the formation of tartronic acid in high yield under acidic conditions, and its disodium salt under basic conditions.[10] Subsequent oxidation of tartronic acid with a Bi–Pt catalyst produces ketomalonic acid. Following this with a multifunctional four-component Ce–Bi–Pt–Pd catalyst, the ketomalonic acid resulting from one-pot glycerol aerobic oxidation is readily polymerized to the corresponding polyether, polyketomalonate, by addition of a cationic or anionic initiator (Figure 8.7).

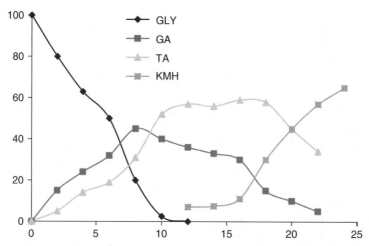

Figure 8.7 One-pot consecutive oxidation of glycerol to ketomalonic (mesoxalic) acid over a multifunctional Ce–Bi–Pt–Pd catalyst. (GLY = glycerol; TA = tartronic acid; GA = glyceric acid; KMH = ketomalonic acid hydrate). (Adapted from Ref. 10, with permission.)

Polyketomalonate, prepared by anionic polymerization in NaOH solution, is an unusual polymer owing to the absence of hydrogen atoms on the polyether carbons. Acidification readily causes decarboxylation to poly(oxymethylene).[10]

Deactivation of the catalyst owing to overoxidation is the most important obstacle to the large-scale application of platinum group metals in liquid-phase oxidation. High temperatures and a low partial pressure of oxygen are required in order to stabilize the catalytically active zero-valent species. Strong adsorption of by-products is also responsible for the deactivation of platinum group metals. Mobility of metal atoms at the catalyst surface causes the metal particles to grow, thus decreasing their specific surface area and the consequent turnover frequency (TOF). The latter phenomenon occurs *via* a reversible metal dissolution–precipitation, favored by strongly coordinated products, whereas irreversible dissolution results in loss of precious catalyst as it is leached into solution.

8.4 Gold and Organocatalysis

Supported gold catalysts are a true second generation class of supported metal catalysts for the oxidation of glycerol. Indeed, great improvements in selectivity and stability have been achieved since the early 2000s. Firstly, the high electrode potential (E° = +1.69 V) of gold is responsible for its well known inertness, which in catalytic terms indicates high

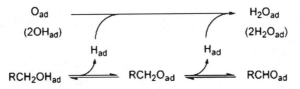

Figure 8.8 Activation of dioxygen and alcohol molecules.

stability, greater resistance to oxygen, and tolerance against poisoning chemical groups such as aliphatic and aromatic amines. A second feature concerns the kinetic aspect of gold catalysis, namely that TOF is strongly related to the size of the metallic gold particles.[14] In particular, a number of investigations into the liquid-phase oxidation of polyols have confirmed that only the small gold particles are catalytically active.[5]

With the new gold-based catalytic systems a valuable new general method for organic synthesis has arrived: primary alcohols and terminal diols can readily be oxidized to carboxylates in aqueous solution in the presence of alkali with almost 100% regioselectivity.[15] Activation of dioxygen and alcohol molecules occurs *via* adsorption on the solid catalyst according to the general pattern shown in Figure 8.8. In the case of polyhydroxylated molecules, the size of particles in the range 2–7 nm correlates linearly to the kinetics, the smaller being the more active, whereas particles larger than 10 nm are virtually inactive.

A great advantage of gold catalysis is the absence of metal over-oxidation and leaching. Kinetic investigation of glycerol oxidation under high oxygen pressure over an Au/C catalyst points to a Langmuir–Hinshelwood mechanism in which the adsorption step of the main compounds, glycerol and glyceric and tartronic acids, is of kinetic relevance.[16]

In the presence of a stoichiometric amount of NaOH (1:1 with respect to the reagent), the aerobic oxidation of glycerol at 60 °C over 1 wt% Au/C under a moderate pressure of oxygen (3–6 bar) affords glycerate with up to 100% selectivity at 54–56 % conversion with a glycerol:Au molar ratio of 538.[17] Newer catalysts containing relatively large particles (>20 nm) show constant selectivity throughout the reaction, giving 92% selectivity to glycerate at full conversion, by oxidizing glycerol at 30 °C with a NaOH : glycerol ratio of 4 and a glycerol : Au molar ratio of 500.[18] Furthermore, a bimetallic Au–Pd/C catalyst enables kinetic control of the oxidative dehydrogenation, producing high selectivity to glyceric acid (69% at 90% conversion), with no overoxidation.[19] Again, relatively large metal particles ensure that glycerate remains stable at higher temperatures, avoiding the formation of tartronate, with a

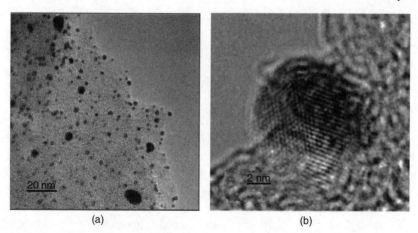

Figure 8.9   (*a*) and (*b*) Au/C catalysts of 6 and 12 nm particle sizes, respectively. (Reproduced from Ref. 20, with permission.)

progressive increment in selectivity towards glycerate as the mean diameter of the particles increases from 2 to 16 nm[20] (Figure 8.9).

Organocatalysis by 2,2,6,6-tetramethylpiperidine-1-oxyl (TEMPO) is a suitable technique for the selective oxidation of glycerol to mesoxalic acid.[21] Coupled to an inexpensive NaClO–Br regenerating oxidant, the catalytic cycle typical of nitroxyl-mediated oxidation selectively produces ketomalonate (98%) along with a small amount of dihydroxy-acetone (2%).

The reaction can conveniently be carried out over microporous sol–gel silica glass doped with the TEMPO radical moiety. This enables rapid separation of the product from the catalyst, which can be recycled (Figure 8.10).

8.5   Electrochemical Oxidation

The catalytic system employing TEMPO as catalyst for the oxidation of glycerol by NaClO can also be used in the electrochemical conversion of glycerol to 1,3-dihydroxyacetone (DHA).[22] Thus, one-pot waste-free oxidation of glycerol to DHA has been achieved by simply applying a small electric potential (1.1 V *vs* Ag–AgCl) to a glycerol solution in water buffered at pH 9.1 in the presence of 15 mol% TEMPO at the surface of a glassy carbon anode. Prolonging the reaction time gives comparable amounts of hydroxypyruvic acid (Figure 8.11).

The catalytic cycle indicated in Figure 8.12 shows how DHA is formed by TEMPO, which is reduced to the hydroxylamine

Figure 8.10 SEM photographs of sol–gel entrapped silica catalysts doped with TEMPO. (Reproduced from Ref. 21, with permission.)

intermediate. The key point of this process is the reoxidation of the latter to TEMPO through parallel electrochemical pathways.

Apart from the oxidation of glycerol to high-value chemicals ranging from DHA to mesoxalic acid, great attention is being paid to the energy produced by direct conversion of glycerol to carbon dioxide and to electrical energy resulting from oxidation in fuel cells. Films of poly-aniline electrodeposited on gold and glassy carbon electrodes[23] and doped with Pt, Pd or Ru,[24] show good activity in the electro-oxidation of glycerol in acidic media. In such partial oxidations, however, the main

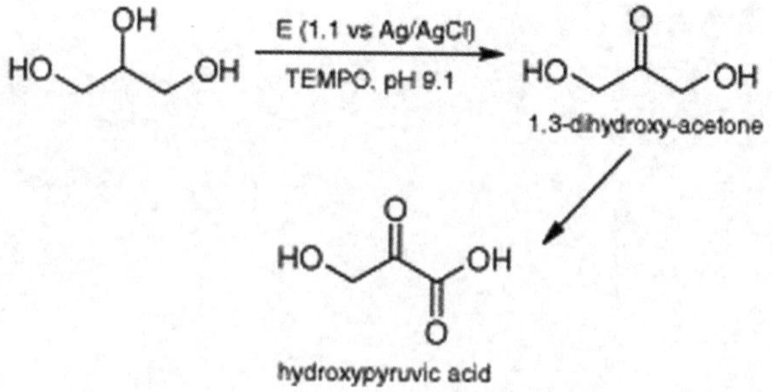

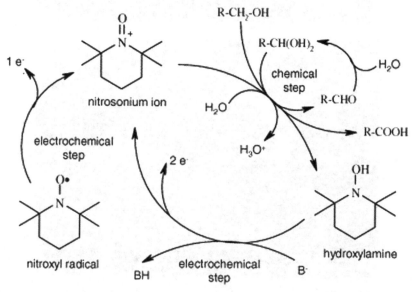

Figure 8.11   Electrochemical oxidation of glycerol to DHA takes place smoothly in aqueous solution. (Reproduced from Ref. 22, with permission.)

Figure 8.12   Electrochemical oxidation of alcohols mediated by TEMPO.

product is glycerate, which results in only 28.6% oxidation of the glycerol fuel and limits the amount of energy produced.

8.6   Biological Oxidation

Industrial fermentation of glycerol has been applied to the synthesis of DHA since the late 1920s. DHA is the main active ingredient in artificial tanning agents,[25] and manufacturers today obtain it by the bioconversion of glycerol extracted from colza or palm tree oil.[26] A variety of micro-organisms and derived enzymes can be used, producing a range of selectivities. Bacterial cells such as *Acetobacter* (*A. suboxydans* ATCC 621, *A. xylinum* A9, *A. kuetzingianus* OUT 8296, *A. pasteurianus* OUT 8299, *A. rancens* OUT 8300, *A. suboxydans* IFO 3254, *A. suboxydans* IFO 3255, *A. suboxydans* IFO 3291 and *A. suboxydans* IFO 3432), *Gluconobacter* (*G. melanogenus* IFO 3293, *G. melanogenus* IFO 3294, *G. capsulatus* IAM 1813, *G. cerinus* IAM 1832, *G. dioxyacelonicus* IAM 1814, *G. glyconicus* IAM 1815 and *G. roseus* IAM 1838), yeasts (*Candida valida* and *Neurospora crassa*) all show high selectivity to DHA.[27] Batch fermentation can be used with *G. oxydans* or *A. suboxydans*. A troublesome feature of microbial fermentation is the tedious inoculation required for every cycle, and improved systems adopt a semi-continuous process in which part of the fermentation broth is allowed to remain in the reactor as the inoculate for the next cycle. Moreover, DHA has an inhibitory effect on bacterial growth and single-stage production restricts DHA concentration to $60\,kg\,m^{-3}$. A two-stage process based on *G. oxidans* has recently been developed to increase DHA productivity,[28] along with a heterogeneous process using cells immobilized on polvinyl alcohol beads.[5] The first stage provides the source of a durable non-product-inhibited culture, and the second stage is a high concentration DHA reaction. Under these conditions a DHA threshold value of $82\,kg\,m^{-3}$ has been reached in the first stage and $161\,kg\,m^{-3}$ in the second reaction.

With a view to providing an outlet for the glycerol by-product of biodiesel, however, the most promising biological oxidation process is the recent employment of glycerol in fuel cells using membrane-immobilized enzymes.[29] Two oxido-reductase enzymes (PQQ-dependent alcohol and aldehyde dehydrogenase, respectively) are immobilized at the surface of a carbon anode in the pores of a Nafion ion exchange copolymer membrane modified with quaternary ammonium groups, in order to expand its pores and make the environment more hydrophobic and enzyme-friendly; this ensures stability of the enzymes for months or even years. These glycerol bioanodes have been incorporated into a glycerol–oxygen biofuel cell which enables multi-step oxidation of glycerol to mesoxalic acid; this utilizes 86% of the energy density of the glycerol and results in power densities of up to $1.21\,mW\,cm^{-2}$ at room temperature. This is very different from metallic electrodes, which give glycerate as the only detectable oxidation product of glycerol, and it shows that the biofuel cell can allow deeper oxidation of the glycerol fuel, increasing overall efficiency and energy density (Figure 8.13).

This technology has been licensed to the company Akemin Inc, and is progressing towards commercialization.[30] The key to commercial development will lie in improving the lifetime and performance of enzymes

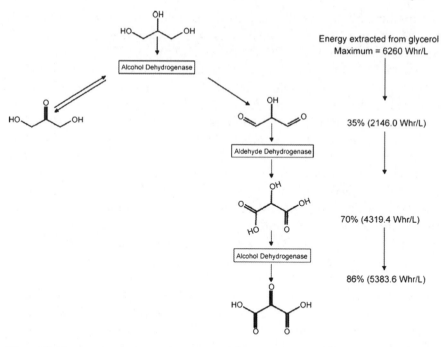

Figure 8.13 Oxidation sequence for glycerol at a PQQ–ADH/PQQ–AldDH-modified bioanode. (Reproduced from Ref. 2, with permission.)

over a range of temperatures. Fuel cells convert the energy stored in the fuel directly into usable electrical power. They are in effect batteries which can be recharged merely by refilling a small fuel tank. One of their main drawbacks is that they are limited to simple fuels such as hydrogen or methanol, each of which gives serious safety concerns—hydrogen is explosive and methanol is flammable and highly toxic. On the other hand glycerol is cheap, readily commercially available, safe, nontoxic and nonflammable, and it contains three times more energy per gallon than liquid hydrogen. Glycerol can also be used in a biofuel cell at 98.9% concentration without damaging the cell, whereas methanol is limited to 40% concentration. As a result, a similar amount of glycerol produces almost four times as much power as methanol, and it is the main alternative being currently considered for portable electronics such as cell phones and laptops. Biofuel cells have the further advantage that they avoid the use of expensive precious metals. On the other hand, their use is restricted by the fact that they can only be used between freezing temperatures and 90 °C, so they are unsuitable for high temperature applications such as automobiles and household generators.

References

1. By the Finnish company Remedal Ltd (www. remedal.com).
2. R. L. Arechederra, B. L. Treu, S. D. Minteer, Development of glycerol–O₂ biofuel cell. *J. Power Sources*, 2007, **173**, 156.
3. T. Mallat and A. Baiker, Oxidation of alcohols with molecular oxygen on solid catalysts. *Chem. Rev.*, 2004, **104**, 3037.
4. L. F. Fieser and M. Fieser, *Advanced Organic Chemistry*, Reinhold, New York, 1961.
5. S. Wei, Q. Song and D. Wei, Repeated use of immobilized *Gluconobacter oxidans* cells for conversion of glycerol to dihydroxyacetone. *Prep. Biochem. Biotechnol.*, 2007, **37**, 67.
6. H. Kimura and K. Tsuto, Catalytic synthesis of DL-serine and glycine from glycerol. *J. Am. Oil Chem. Soc.*, 1993, **70**, 1027.
7. F. Dobinson, Ozonization of malonic acid in aqueous solution. *Chem. Ind. Lond.*, 1959, **26**, 853.
8. W. R. Davis, J. Tomsho, S. Nikam, E. M. Cook, D. Somand and J. A. Peliska, Inhibition of HIV-1 reverse transcriptase-catalyzed DNA strand transfer reactions by 4-chlorophenylhydrazone of mesoxalic acid. *Biochemistry*, 2000, **39**, 14279.
9. The Merck Index, 14th edn, New York: Merck and Co, 2006.
10. H. Kimura, Oxidation assisted new reaction of glycerol. *Polym. Adv. Technol.*, 2001, **12**, 697.
11. H. Kimura, Selective oxidation of glycerol on a platinum–bismuth catalyst by using a fixed bed reactor. *Appl. Catal., A*, 1993, **105**, 147.
12. P. Fordham, M. Besson and P. Gallezot, Catalytic oxidation with air of tartronic acid to mesoxalic acid on bismuth-promoted platinum. *Catal. Lett.*, 1997, **46**, 195.
13. P. Fordham, R. Garcia, M. Besson, P. Gallezot, Selective oxidation with air of glyceric acid to hydroxypyruvic acid and tartronic to mesoxalic acid on PtBi/C, In H. U. Blaser, A. Baiker, R. Prins, Heterogenous catalysis and fine chemicals IV, *Stud. Surf. Sci. Catal.*, 1997, **108**, 429; Elsevier: Amsterdam.
14. G. C. Bond, C. Louis and D. T. Thomson, *Catalysis by Gold*, London: Imperial College Press, 2006.
15. M. Comotti, C. Della Pina, R. Matarrese and M. Rossi, The catalytic activity of naked gold particles. *Angew. Chem., Int. Ed.*, 2004, **43**, 5812.
16. S. Demirel, M. Lucas, J. Warna, T. Salmi, D. Murzin and P. Claus, Reaction kinetics and modelling of the gold catalyzed glycerol oxidation. *Top. Catal.*, 2007, **44**, 299.

17. S. Carrettin, P. McMorn, P. Johnston, K. Griffin, C. J. Kiely, G. A. Attard and G. J. Hutchings, Oxidation of glycerol using supported gold catalysts. *Top. Catal.*, 2004, **27**, 131.

18. F. Porta and L. Prati, Selective oxidation of glycerol to sodium glycerate with gold-on-carbon catalyst: an insight into reaction selectivity. *J. Catal.*, 2004, **224**, 397.

19. L. Prati, A. Villa, F. Porta, D. Wang and D. Su, Single-phase gold/palladium catalyst: The nature of synergistic effect. *Catal. Today*, 2007, **122**, 386.

20. N. Dimitratos, J. A. Lopez-Sanchez, D. Lennon, F. Porta, L. Prati and A. Villa, Effect of particle size on monometallic and bimetallic (Au,Pd)/C on the liquid phase oxidation of glycerol. *Catal. Lett.*, 2006, **108**, 147.

21. R. Ciriminna and M. Pagliaro, One-pot homogeneous and heterogeneous oxidation of glycerol to ketomalonic acid mediated by TEMPO. *Adv. Synth. Catal.*, 2003, **345**, 383.

22. R. Ciriminna, G. Palmisano, C. Della Pina, M. Rossi and M. Pagliaro, One-pot electrocatalytic oxidation of glycerol to DHA. *Tetrahedron Lett.*, 2006, **47**, 6993.

23. E. C. Venancio, W. T. Napporn and A. J. Motheo, Electrooxidation of glycerol on platinum dispersed in polyaniline matrices. *Electrochim. Acta*, 2002, **47**, 1495.

24. A. Nirmala Grace and K. Pandian, Pt, Pt–Pd and Pt–Pd/Ru nanoparticles entrapped polyaniline electrodes: A potent electrocatalyst towards the oxidation of glycerol. *Electroch. Comm.*, 2006, **8**, 1340.

25. R. K. Chaudhuri, Dihydroxyacetone: chemistry and applications, in *Self-tanning products, chemistry and manufacture of cosmetics*, vol. 3, ed. M. L. Schlossman, Weimar, Texas: C.H.I.P.S. Books, 3rd edn, 2002.

26. For instance, by Soliance in France (www.groupesoliance.com).

27. K. Nabe, N. Izuo, S. Yamada and I. Chibata, Conversion of glycerol to dihydroxyacetone by immobilized whole cells of *Acetobacter xylinum*. *Appl. Environ. Microbiol.*, 1979, **38**, 1056.

28. R. Bauer, N. Katsikis, S. Varga and D. Hekmat, Study of the inhibitory effect of the product dihydroxyacetone on *Gluconobacter oxidans* in a semi-continuous two-stage repeated-fed-batch process. *Bioprocess Biosyst. Engin.*, 2005, **28**, 37.

29. The original patent describing the technology is: N. L. Akers, C. M. Moore, S. D. Minteer, Enzyme immobilization for use in biofuel cells and sensors, US2004101741.

30. According to the inventor it will take about 3–5 years before glycerol clips could be used to power mobile phones (Prof S. F. Minteers, *personal communication to M. P.*).

CHAPTER 9
Additives for Cement

9.1 Polyols as Additives for Cement

Construction materials for the manufacture of buildings, roads and bridges constitute one of the largest volume markets in human activity. Hydraulic cements, for instance, are employed annually in quantities extending to billions of tonnes and global consumption overall is expanding rapidly, particularly in China, India and other Far Eastern regions (Figure 9.1).

Hydraulic cements are produced by calcining a mixture of calcareous and argillaceous materials comprising limestone, sand, shale, clay and iron ore to produce a sintered "clinker". Prior to calcination the raw materials are crushed to below 10 cm in size. In either wet or dry production processes the quantity of each raw material is adjusted to give the required chemical composition and fed to a rotating ball mill, with or without water. In the resulting material the majority of particles are below 75 µm (Figure 9.2). The material discharged from the mill is known as "kiln feed". The chemical reactions after evaporating all moisture include calcining the limestone to produce free calcium oxide and reacting the calcium oxide with the sand, shale, clay and iron. Calcining is carried out in an intense flame formed by burning coal or pet coke, and results in the black nodular clinker. The clinker is ground by ball milling along with gypsum and other additives to form the cement. The fineness of the final product and the quantity of gypsum and other materials added can be varied to give the required performance for each individual product.

Among hydraulic cements the Portland types are by far the most important in terms of production volume. In this case the clinker is mixed with the required quantity of gypsum and ground, usually in a ball mill, to a finely divided state with a large surface area to produce the finished cement. Clearly this grinding process consumes a considerable amount of

RSC Green Chemistry Book Series
The Future of Glycerol: New Uses of a Versatile Raw Material
By Mario Pagliaro and Michele Rossi
© Mario Pagliaro and Michele Rossi 2008

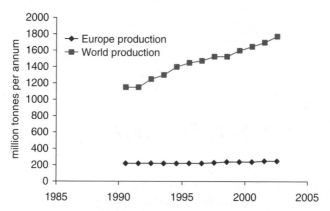

Figure 9.1   World consumption of cement is rapidly on the rise, whereas in Europe it
is constant at about 200 million tonnes per annum.
(Source: Grace Construction Products.)

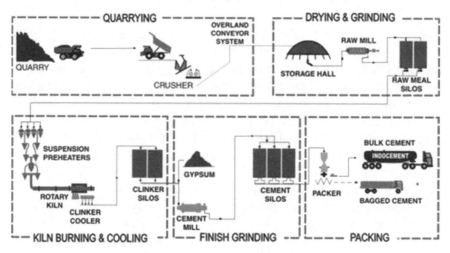

Figure 9.2   Scheme of typical dry process for cement manufacture.
(Source: Indocement.)

energy, in fact energy costs form 40% of the total cost of production. Of
this 80% is accounted for by the calcination and grinding processes. Heat
for calcination is normally generated by burning pet coke, but the mechanical and thermal energy for the cooler is provided by electricity, amount-
ing to around 20% of overall energy consumption (Table 9.1). While
consumption of thermal energy has fallen almost to the technical minimum of 3500 MJ per tonne of clinker (Figure 9.3), the electricity required
is on the increase due to increasing demand for higher grade cements.

A common practice in the cement industry is to add small amounts of
grinding additives during milling to lower the friction between particles.

Table 9.1 Electrical energy represents about 20% of the overall energy input for cement manufacture, with typical values in the range 90–130 KWh per tonne.

	kWh per tonne
Crushing	1–3
Grinding of the crude mixture	8–30
Homogenization of the crude	1–2
Cooking	16–18
Grinding finished cement	30–80

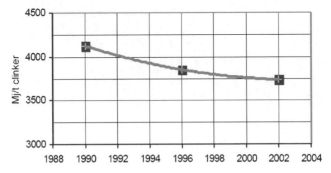

Figure 9.3 Evolution of specific thermal energy consumption for the production of 1 tonne of clinker in Italy. Similar trends are found in Germany and other leading industralized countries.
(Source: Grace Construction Products.)

This has the effect of reducing the energy needed to grind the clinker to the required particle size (Figure 9.4).

An important problem in the handling of cement-like materials is in the packaging of the powdered product. After grinding most cements become semi-rigid when vibrated and compacted and will not flow under a consistent mechanical effort. Lowering the "pack set inhibition", the energy required to initiate flow, is of paramount importance in discharging cement powder from storage silos and trucks.

Water soluble polyols, mainly glycols and polyglycerols,[1] and the acetic acid esters of glycerol,[2] are useful cement additives which act as grinding aids and reduce pack set inhibition.[3] Grinding aids reduce the cost of cement manufacture by decreasing the attractive forces between the cement particles which cause agglomeration. They are added to the cement during the grinding process in the proportion 150–500 g per tonne of cement, resulting in:

- increased cement flowability and reduced pack set;
- increased grinding efficiency and mill output;

Figure 9.4   A ball mill used to grind clinker and produce cement.
(Photo courtesy of Grace Construction Products.)

- reduced unit power costs; and
- reduced handling and pumping costs.

All the chemicals which are found to improve the grinding process are strongly polar in nature. Their effect results from decreasing the causes of agglomeration. The grinding aids are adsorbed on the newly created surface of the particles, neutralising electrical charges and reducing their tendency to reagglomerate (Figure 9.5). As a result the cement grains become easier to grind into finer particles.

Apart from polyols, other types of additives may be used, including water-soluble aliphatic alkali carboxylates,[3] nitrogen-containing hydroxyl compounds such as triethanolamine,[4] and sulphonated lignins.[5] A combination of more than one component is frequent employed.[6]

A third issue needing to be overcome by cement additives is the surface cracking of Portland cement, which appears under the effects of wind, low humidity and high temperature. An additional factor improved by additives is compression strength. Compression strength is the capability of a manufactured cement article to withstand pressure. When the critical compression strength is reached, fractures are generated at the surface which may cause the article to break. Since high compression strength is so important, a variety of additives have been developed to improve this property. These are preferably added at the clinker milling stage.

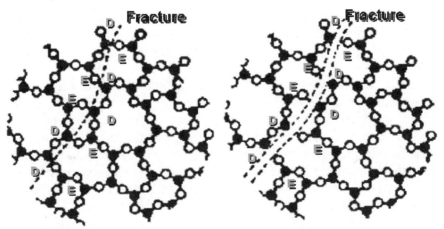

Figure 9.5 Silica crystal structure. Units D and E refer, respectively, to oxygen-deficient ($+$) and oxygen-excess ($-$) sites.

9.2 Glycerol as an Anticracking and Waterproofing Agent

Pure glycerol gives good results in improving compression strength, but its industrial usage has in the past been hindered by the high cost of the pure material. Glycerol has also been employed in conjunction with film-forming polymeric materials for preventing surface cracking.[7] A continuous coating comprising an aqueous alcoholic solution of glycerol or a polyglycerol, or a mixture of the two, is applied to a fresh uncured concrete surface containing a film-forming polymeric modifier such as a styrene butadiene copolymer. By comparison with a blank experiment, no cracking is observed in a test specimen.

Small quantities of glycerol can also be employed in an additive to improve the waterproofing of concrete.[8] A typical additive has a number of components, for example alumina (10%), alkali (4%), fatty acid (3.5%) and glycerol (0.5%). An aqueous slurry of this composition can be added to fresh cement used for repairing old concrete structures.

9.3 Raw Glycerol as a Quality Enhancer and Grinding Aid

Crude glycerol arising from biodiesel production is an excellent quality enhancer for cement, enhancing its resistance to compression.[9] This discovery, resulting from collaboration between the University of Milan, the National Research Council of Italy and Grace Construction Products, opens up a considerable market for biodiesel-generated glycerol. Furthermore, since biodiesel production sites are generally not far distant from the large variety of cement production sites in every continent, cement manufacturers can rely on a convenient and abundant source of glycerol for improving the mechanical properties of their products.

Additives composed primarily of organic materials with only minor proportions of inorganic ingredients are now used in 60% of world cement production. The organic components are derived from ethylene oxide, and thus from petroleum. Table 9.2 shows the results of industrial tests carried out in Belgium, Italy and Greece on three types of clinker, the cement precursor blended with gypsum to produce cement, to which raw and pure glycerol from biodiesel production has been added in amounts comparable to those used with traditional additives.

The tests were carried out according to international standards and the final concentration of additive, as a 50% solution in water, was usually in the range 200–400 ppm. In all cases glycerol in the crude form, containing 5% NaCl and with a color approaching that of lager (due to the presence of natural dyes) was found to produce similar or improved mechanical and chemical properties to those produced by three other

Table 9.2 Granulometry and performance properties of three cement clinkers with addition of pure or raw glycerol.

Cement			Laser PSD[a]				Compressive strength (MPa)			
Additive	ppm	Blaine[b] (cm^2 g^{-1})	R32 (%)	R45 (%)	R63 (%)	R90 (%)	1d	2d	7d	28d
Belgium										
Blank	–	3230	21.9	10.5	3.2	0.1	–	25.9	45.0	58.2
Pure glycerol 50%	400	3290	24.7	12.5	4.1	0.4	–	26.7	45.0	58.8
Raw glycerol 50%	400	3160	26.5	14.2	5.2	0.7	–	**28.3**	**46.9**	**60.8**
Greece										
Blank	–	3570	18.8	8.7	2.6	0.2	16.1	–	41.6	53.0
Pure glycerol 50%	400	3550	22.1	10.8	3.4	0.3	20.1	–	41.6	53.8
Raw glycerol 50%	400	3590	21.5	10.7	3.6	0.5	18.1	–	**43.8**	**56.6**
Italy										
Blank	–	3560	27.5	16.4	7.8	2.4	–	22.2	38.0	52.9
Pure glycerol 50%	400	3480	33.6	21.1	10.5	3.2	–	26.3	39.0	49.6
Raw glycerol 50%	400	3590	33.0	21.3	11.1	4.0	–	24.8	**40.8**	51.8

[a]Laser PSD: granulometry distribution (number of particles with diameter greater a certain figure, in this case: 32, 45, 63 and 90 μm);
[b]Blaine: measure of cement fineness.

additives, including pure glycerol itself. This is surprising, and points to synergy between glycerol and the inorganic salts present. According to industry experts, initial market requirement of crude glycerol is estimated to be about 1000 tonnes per annum, roughly 2% of Europe's 50 000 tonne market for glycerol as a cement additive.

A second major benefit of glycerol as a cement additive arises from its use in grinding technology. Use of low levels of glycerol, either in pure or raw form as a by-product of biodiesel production, gives significant improvements in mechanical properties in the milling of a range of different doped cement clinkers. A similar or better performance is obtained in comparison to use of the more expensive triethanolamine (TEA) and diethylene glycol (DEG), typically used as high value cement performance enhancers (Table 9.3).

In cement production the energetic balance is dominated by thermal heating of the furnace and the electrical energy required by the mill. As mentioned earlier, these factors account for as much as 40% of total production costs (fuel 3500 MJ per tonne of clinker and electricity 90–130 KWh per tonne of cement). The benefit of grinding additives is in decreasing electrical energy consumption by about 10%, which is a remarkable result.[10] The annual market for grinding aids in continental Europe, not including performance- and quality-enhancing additives, is

Table 9.3 Properties of clinkers with addition of raw glycerol (GLY), triethanolamine (TEA) or diethylene glycol (DEG). Passing refers to percentage of particles passing through a standard sieve.

| Sample | Particle size (μm) | | | | Blaine ($cm^2 g^{-1}$) | RPM (s^{-1}) | Packset (s) |
	32.0	45.0	63.0	90.0			
Clinker Italy A							
Passing% –		74.8	85.6	94.3	99.0	3620	3200
Passing% TEA		74.4	86.4	95.2	99.4	3620	2500
Passing% DEG		74.3	85.9	94.7	99.2	3550	2500
Passing% GLY		72.0	84.1	93.6	98.9	3720	2500
Clinker Italy B							
Passing% –	71.2	82.4	91.7	97.8	3580	2200	41
Passing% TEA	73.8	85.3	93.8	98.6	3510	2250	29
Passing% DEG	74.1	84.8	93.3	98.5	3590	2200	28
Passing% GLY	79.9	89.8	96.3	99.5	3590	2050	32
Clinker Greece							
Passing% –	80.0	89.7	96.3	99.5	3610	2350	68
Passing% TEA	77.5	88.1	95.4	99.3	3510	2400	21
Passing% DEG	78.5	89.0	96.0	99.4	3620	2200	38
Passing% GLY	77.3	88.1	95.5	99.3	3530	2400	32

25 000 tonnes, only 0.012% of European cement production which totals 200 million tonnes.

References

1. H. H. Moorer and C. M. Anderegg, Cement grinding aid and set retarder, US4204877 (1980).
2. I. C. Bechtold, Portland cement and its manufacture, US2225146 (1940).
3. L. A. Jardine, J. H. Cheung and W. M. Freitas, High early strength cement and additives and methods for making the same, US6641661 (2003).
4. G. R. Tucker, C. W. Tucker, H. L. Kennedy, M. S. Renner, Concrete and hydraulic cement, US2031621 (1936).
5. J. G. Mark, Concrete and hydraulic cement, US2141570 (1938).
6. H. H. Moorer and C. M. Anderegg, Cement grinding aid and pack set inhibitor, US3615785 (1971).
7. T. Y. Moon and R. H. Cooper, Method of preventing surface cracking of Portland cement mortar and concrete containing a film forming polymer modifier, US4141737 (1979).
8. L. Ehrenburg, Waterproofing agents for cement and concrete, US3047407 (1962).
9. M. Rossi, M. Pagliaro, R. Ciriminna, C. della Pina, W. Kesber and P. Forni, Improved compression strength cement, WO2006051574.
10. P. Forni, M. Rossi and M. Pagliaro, *unpublished results*.

CHAPTER 10
Sustainability of Bioglycerol

10.1 Biofuels: A Triple Bottom Line Analysis

To be a viable alternative to conventional fuels, a biofuel should provide a net energy gain, offer environmental benefits, be economically competitive, and available in large quantities without reducing food supplies.[1] Brundtland's definition of sustainability – safeguarding the world for future generations – has three elements: social, economic and environmental. All three aspects must be satisfied if something is to be considered truly sustainable. In other words, anything which protects the environment by driving people into poverty, or which supports the economy while undermining environmental services, will inevitably collapse, often with devastating consequences for people either living today or for future generations. Ecosystems, including those involving humans, have maintained themselves for tens of thousands of years and future business should manage and measure its success not just in terms of profit, but rather by a "triple bottom line" of economic, social and environmental performance.[2]

Evaluating the economics of biofuels using full life-cycle accounting for biodiesel from soybeans and ethanol from corn grain, biodiesel yields 93% more energy than the energy invested in its production, while ethanol yields only 25% more.[3] This advantage of biodiesel over ethanol comes from lower agricultural input and more efficient conversion of feedstocks to fuel. Hence, compared with ethanol, biodiesel releases just 1.0%, 8.3%, and 13% as much agricultural nitrogen, phosphorus, and pesticide pollutant, respectively, per net energy gain. Biodiesel also releases less air pollutant per net energy gain than does ethanol. Thus, relative to the fossil fuels they displace, greenhouse gas emissions are reduced 41% by the production and combustion of biodiesel, compared with 12% for ethanol. The results of a study (Figure 10.1), which

RSC Green Chemistry Book Series
The Future of Glycerol: New Uses of a Versatile Raw Material
By Mario Pagliaro and Michele Rossi
© Mario Pagliaro and Michele Rossi 2008

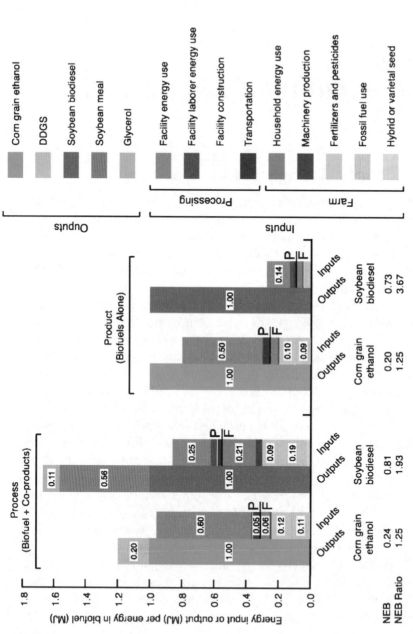

Figure 10.1 NEB (net energy balance) of corn grain ethanol and soybean biodiesel production. Energy inputs (nine categories) are consistently ordered in each set of inputs, as in the legend. Individual inputs and outputs ≥0.05 are labeled. The NEB (energy output − energy input) and NEB ratio (energy output/energy input) for each biofuel are presented, both for the entire production process (*left*) and for the biofuel only (*i.e.*, after excluding coproduct energy credits and energy allocated to coproduct production) (*right*). (Reproduced from Ref. 3, with permission.)

includes the energy costs of farm machinery and processing facilities, together with recent advances in crop yields and biofuel production efficiencies, refute previous assertions[4] that biofuel requires more energy to make than it yields.

Indeed, the net energy balance (NEB) advantages of soybean biodiesel are confirmed using five different methods of accounting for the energy credits of the co-products. Thus biodiesel offers sufficient environmental advantages to merit a subsidy given to otherwise economically uncompetitive biofuels which are justified by their reduced life-cycle environmental impact.

Governments in Europe and the USA provide a subsidy in the form of tax exemption for each litre of biodiesel produced.[5] Furthermore, both in the EU and the USA ethanol and biodiesel producers also benefit from federal crop subsidies which give lower soybean prices. The main logic in subsidizing the biodiesel industry is to support the farmer, reduce unemployment, prevent neglect of the countryside and, in democracies, to win the votes of the beneficiaries. Incidentally, it may be worth noting here that a strain of *Escherichia coli* bacterium (Figure 10.2) could potentially ferment glycerol anaerobically into ethanol and valuable green chemicals, with an estimated operational cost 40% lower than first-generation ethanol production from corn.[6]

Because fossil energy use includes environmental costs not captured in market prices, whether a biofuel provides a net benefit to society

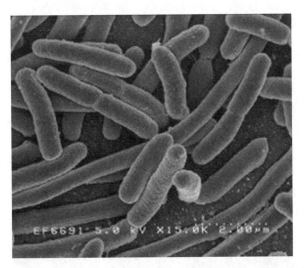

Figure 10.2   Ubiquitous *E. coli* bacteria. These bacteria are widely used in industrial fermentation processes and can efficiently convert glycerol to ethanol. (Photo courtesy of Wikipedia.org.)

depends not only on whether it is cost competitive but also on its environmental costs and benefits *versus* fossil fuel alternatives. Demand,
especially for ethanol, mostly arises from laws and regulations man-
dating the blending of biofuels in some proportion into petroleum and
diesel. Moreover, further increases in petroleum prices above the 2007
average improve the cost competitiveness of biofuels.

As crude oil prices have skyrocketed since the late 1990s (current price
around US$100 per barrel, from US$20 per barrel in 1998), so has the
interest in biodiesel production and use. For example, in Colombia—the
largest palm-oil producer in the Americas—by 2003 oil palm plantations
had occupied 188 000 hectares and, including fields planted but not yet
producing, the total is now closer to 300 000 hectares. Palm oil used to
be used entirely for cooking and soaps but 35% of output is now used
for biodiesel fuel. The same trend is taking place in Malaysia, Argentina
and other developing countries. The US production of biodiesel is 30–40
million US gallons and is expected to grow at a rate of 50–80% per year,
with a probable figure of 400 million gallons by the year 2012.

Chemical and petrochemical companies, too, are starting to invest
heavily in biofuels. Recently published estimates predict that annual demand for biodiesel will grow within the next few years from 6 to 9 million
tonnes in the USA and from 5 to 14 million tonnes in the European
Union. As a representative current example, the British company D1–BP
Fuel Crops (a joint venture between BP and D1 Oils) will invest US$160
million over the next five years to cultivate *Jatropha curcas* (Figure 10.3)
as a biodiesel feedstock in Southeast Asia, Southern Africa, Central and
South America, and India. The plant is a drought-resistant, oilseed-
bearing tree and does not compete with food crops for good agricultural

Figure 10.3   *Jatropha curcas* oilseeds.
              (Photo courtesy: Dreamstime.)

land, nor does it adversely impact the rainforest, because *Jatropha* can be grown on land where other crops struggle, and has the further benefit of requiring less irrigation. Moreover, inedible oils are not subject to the same market competitive demand pressures as food oils.

However, no biofuel can replace petroleum to any extent without impacting on food supplies. Even dedicating the entire US corn and soybean production to biofuels would meet only 12% of petroleum and 6% of diesel demand. Until the recent increases in petroleum prices, high production costs made biofuels unprofitable unless subsidised. Transportation biofuels such as synfuel hydrocarbons or cellulosic ethanol, if produced from low-input biomass grown on agriculturally marginal land or from waste biomass, will provide much more plentiful supplies and environmental benefits than food-based biofuels. Results of a comprehensive study[1] reveal that biofuels could provide 37% of US transport fuel within the next 25 years, and that biofuels could replace 20–30% of the oil used in European Union countries during this time frame.

10.2 Bioglycerol Economics

Global production of bioglycerol from biodiesel (Figure 10.4) has climbed from 200 000 tonnes in 1995 to 600 000 tonnes in 2006. This glycerol stream typically contains a mixture of glycerol itself, with methanol, water, inorganic salts (catalyst residues), free fatty acids, unreacted mono-, di-, and triglycerides, methyl esters, and a variety of

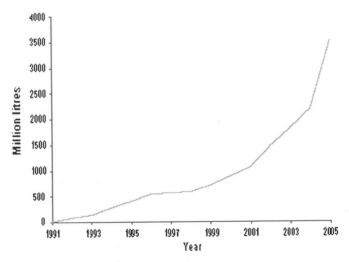

Figure 10.4 Global glycerol production (1975–2005).
(Source: F O Licht.)

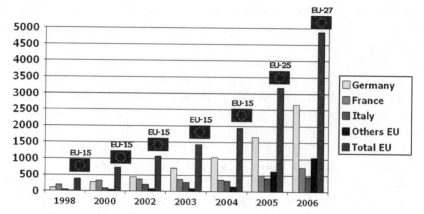

Figure 10.5 EU member states' biodiesel production (1998–2006).
(Source: European Biodiesel Board.)

other "matter organic non-glycerol" (MONG) in varying amounts. Most of the growth has come within the last three years, due to glycerol arising as a by-product of the production of biodiesel.[7] In the same period, the spot market price for refined glycerol has fallen from €1500 to €330 per tonne. Figure 10.4 shows that world production of biodiesel quadrupled between 2000 and 2005 and so, correspondingly, has global glycerol production.[8]

In addition, Figure 10.5 shows that among the 27 member states of the European Union alone, total biodiesel production for 2006 was in excess of 4.8 million tonnes, an increase of 54% on 2005, which is equivalent to 480 000 tonnes of crude glycerol.

In this context, the alternative process for biodiesel production based on the high-pressure hydrogenation of fatty acids might alter the availability of crude glycerol and thus affect the trend shown in Figure 10.4.[9] However, even though the product outperforms biodiesel produced by conventional transesterification, the high cost of the investment required for the high-pressure unit (approximately €200 million for the two units being built at Porvoo refinery in Finland, Figure 10.6) implies that both biodiesel fuels will remain available on the market in the coming years.

A model based on the production and sale of 80% glycerol by mass, and assigning it a value of US$0.33 per kg (consistent with recent prices), predicts an inverse linear relationship between the production cost of biodiesel and variations in the market price of glycerol, with an increase of US$0.0022 per litre for each US$0.022 reduction in glycerol price per kg. Hence, in order to expand biodiesel production, economical utilization pathways for the glycerol by-product *must* be developed,

Figure 10.6 The first NExBTL production plant is currently under construction at Neste Oil's Porvoo refinery with a rated annual capacity of 170 000 tonnes. It came on stream in summer 2007.
(Photo courtesy of Suomen Ilmakuva Oy.)

either by expanding the existing markets for glycerol or, better still, by establishing new marketable value-added derivatives. Developing new industrial uses for glycerol will greatly increase the net energy and sustainability of biodiesel.

To produce USP grade glycerol, the crude material from biodiesel production must be purified. Cargill, for instance, has built its own refinery in conjunction with its 37.5 million gallon per annum biodiesel plant in Iowa, making it the first company in North America to combine soybean crushing, biodiesel and USP grade glycerol production in a single plant. However, refining crude glycerol is too expensive for small-scale producers, who need a price rise to sell the crude which they currently have to give away just to get it out of their storage tanks. In the USA, for instance, up to mid-2007 it was common practise to market 50% of total crude glycerol and sell the remainder at minimal price.

The impact of biodiesel production in the EU on domestic glycerol supplies has been small in comparison to the effect of imports, which doubled again in 2006. In Europe an EU directive adopted in 2003 aims to promote the use of transport fuels made from biomass, as well as other renewable fuels, and requires that by the year 2010, some 5.75% of the total fuel consumed in the EU must come from renewable sources. Since directives are binding on EU member states this has naturally provided a sharp incentive for producing biodiesel, especially so in France and Germany. Over the past decade environmental legislation has progressively lowered the sulfur content in fossil diesel fuel, and biodiesel is widely used in the EU as a fuel additive to achieve this objective; it also provides excellent lubricant properties. Finally, since the middle of the 1990s production of natural fatty alcohols in Asia has also increased steadily, adding to the glycerol surplus. For instance countries such as Malaysia and Indonesia are major producers of glycerol, mostly derived from palm oil, and sell it to China where demand for this chemical is burgeoning. As a result the global market for glycerol reached 900 000 tonnes in 2006, of which 600 000 tonnes came from biodiesel production, compared to only 60 000 tonnes in 2001. Such a rapid surplus of refined glycerol soon caused prices to plummet to US$0.33 per kg or below.

Having become part of the fuel market, glycerol is now an important element of the global market scene.[10] The graph in Figure 10.7 shows the price trend for refined glycerol up to 2003 and reports the prediction by a US researcher that by 2010 the glycerol price will fall to as low as US$0.35 per lb.[11]

Historically the price of glycerol has been dictated primarily by factors as diverse as weather, regulatory costs, plant shut-downs and politics. Over the past 30 years glycerol prices have oscillated in roller-coaster fashion from the low $0.20s per lb to a record $1.20 in 2001 (Figure 10.8).

Yet subsequent events have shown that the above prediction was conservative. The glycerol price has plunged rapidly and since the first

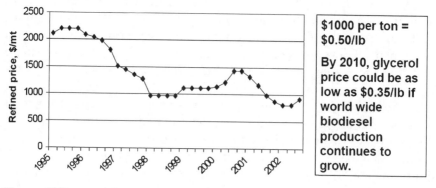

Figure 10.7 Price trend of industrial glycerol. (Reproduced from Ref. 11, with permission.)

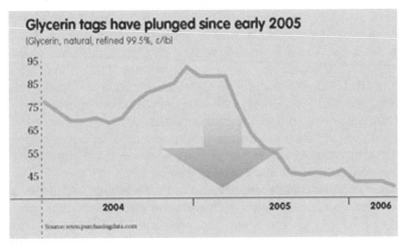

Figure 10.8 Price trend of refined glycerol. (Reproduced from Ref. 12, with permission.)

quarter of 2005 it has been selling for US$0.30–0.55 per lb, depending on the grade, whereas during 2006 the worldwide glut of glycerol created from biodiesel production drove the price below 2005 levels.[12] The market is, however, self-correcting. In 2005, a comprehensive market study[13] predicted that glycerol prices in the long-term (2008) would not fall below €400 per tonne in Europe and $0.30 per lb in the USA. The reasoning was that due to its consistently low price glycerol usage could be expected to expand, especially in replacing propylene glycol, whose price since the US hurricanes of September 2005 had gone up by a factor of six. This has in fact proven to be the case. Already in 2006 purchasers of vegetable-based refined glycerol could obtain contracts in the range US$0.38–0.42 per lb, and by mid-2007 were paying "in the high 40s".[14]

10.3 Glycerol: A Platform Chemical for the Biorefinery

Glycerol stands to develop into an even more commonly used chemical than it is at present. It will become a substitute for many common petrochemicals on the market or, to express it in the words of one practitioner, "glycerol may become the next biodiesel". Globally, glycerol arising from biodiesel production has increased from 200 000 tonnes in 2004 to 800 000 tonnes in 2007. Well-regarded experts in the industry consider that biodiesel production could rise to 9 million tonnes by 2010 in Europe alone, translating into 900 000 tonnes of glycerol.[13] This gives a clear indication of the developing status of glycerol as a key raw material. In three to five years it will be seen as an environmentally friendly way of replacing other competing petroleum products, and it is one of the next chemical platforms which will become widely available. As examples, in the USA Dow, Senergy Chemical, ADM, Cargill and Ashland are all set to produce glycerol-derived industrial products such as propylene and ethylene glycols. In Europe, Solvay is already making epichlorohydrin from biodiesel glycerol and will start large-scale production in Asia within two years. Along with Dow, Grace in its turn will commercialize large quantities of concrete with crude glycerol addition.

Experts realistically predict the end of cheap oil by 2040 at the latest, since increased consumption will irrevocably diminish fossil raw materials and build up environmental pressure.[15] It follows that a progressive move by the chemical industry towards renewable feedstocks will become an inescapable necessity. We can already witness this development as chemical manufacturers come to terms with the rising cost of oil and natural gas. The transition to a more bio-based production system is now underway. In the near future, biorefineries (Figure 10.9) in which biomass is catalytically converted to pharmaceuticals, agricultural chemicals, plastics and transportation fuels will take the place of petrochemical plants using chemicals derived from fossil resources.[16]

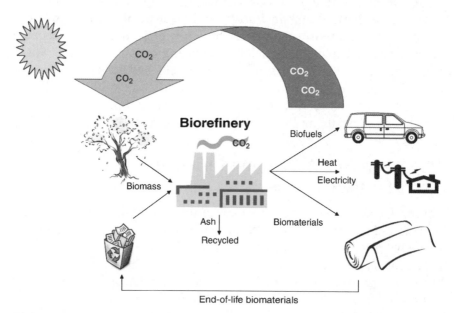

Figure 10.9   The fully integrated agro–biofuel–biomaterial–biopower cycle for sustainable technologies. (Reproduced from Ref. 16, with permission.)

A biorefinery is a facility which uses biomass conversion processes and equipment to produce fuel, power and chemicals from biomass. In concept it is analogous to today's petroleum refineries, which produce a range of fuels and other products from petroleum. Conversion of renewable materials to hydrogen assists in utilization of renewable energy sources, and conversion to commodity chemicals facilitates the replacement of petroleum by renewable sources. Glycerol will become a key renewable feedstock for future biorefineries in which a number of commodity chemicals (Figure 10.10) will be derived from it, especially now that an energy- and atom-efficient conversion of glycerol to syngas has been developed.[17]

Focusing on recent developments in new applications and conversion of glycerol into value-added chemicals (Figure 10.10), we have shown in this book how glycerol is becoming a key raw material for future biorefineries. Human chemical ingenuity has rapidly opened the route to the creation of glycerol derivatives that are finding use in fields as diverse as fuels, chemicals, and the automotive, pharmaceutical, detergent and building industries.

As China and Islamic finance rapidly become central players in the global economy, the value of intellectual property will be progressively weakened.[18] It is now markets, rather than ideas, which generate products. This means that successful products are increasingly copied or developed, and the original product loses its unique status. This will have a considerable impact on the chemical production of biofuels, which will increasingly be produced in developing countries. The technology to make biodiesel is simple and, bearing in mind the new heterogeneous processes, it has reached a high degree of efficiency. The pure glycerol generated as a by-product will further lower the price, including that of USP-grade glycerol, and this is before taking account of the surplus crude glycerol currently being given away free. The raw materials are not localized in a few countries, but instead their production is increasingly determined by land availability. In today's energy-eager world, the price of oil is high (US$100 per barrel as we write) and can only increase, both in the medium and long term. All this makes the production of biofuel an inevitable reality, even beyond the provision of governmental subsidies. The overall consequence is that glycerol will become a central raw material for the chemical industry, along with interesting novelties that, we argue, will originate in Latin America, South-East Asia and Africa. It will be interesting to cover developments in future editions of this book.

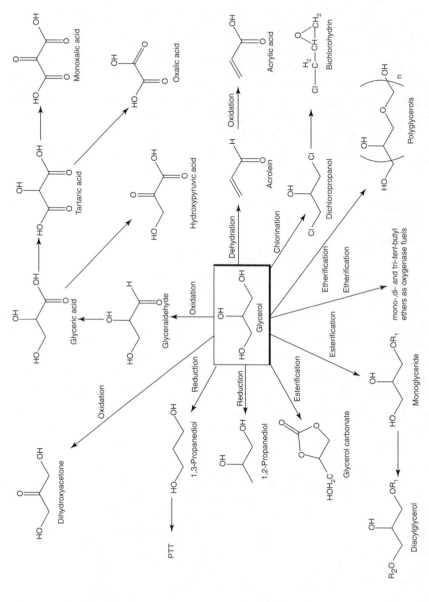

Figure 10.10 Commodity chemicals from glycerol.

References

1. See the report *Biofuels for transportation: global potential and implications for sustainable agriculture and energy in the 21st Century* by the Worldwatch Institute, the German Agencies for Technical Cooperation (GTZ) and Renewable Resources (FNR), 2006.
2. D. Brown, J. Dillard, R. S. Marshall, Triple Bottom Line: A business metaphor for a social construct, Portland State University, School of Business Administration, 2006 (www.recercat.net/bitstream/2072/2223/1/UABDT06-2.pdf).
3. J. Hill, E. Nelson, D. Tilman, S. Polasky and D. Tiffany, Environmental, economic, and energetic costs and benefits of biodiesel and ethanol biofuels. *Proc. Natl. Acad. Sci.* 2006, **103**, 1206. The NEB for corn grain ethanol is small, providing 25% more energy than is required for its production, mostly due to the high energy input required to produce corn and to convert it into ethanol. Almost all of the NEB is attributable to the energy credit for its dry grain with solubles (DDGS) co-product, which is animal feed.
4. Biodiesel from soybean would require 27% more energy than is required in its production, and from sunflower even 118% more: D. Pimentel, T. W. Patzek, Ethanol production using corn, switchgrass, and wood; biodiesel production using soybean and sunflower. *Nat. Resources Res.* 2005, **14**, 65.
5. A. Demirbas, *Energ. Source*, 2003, **25**, 457. A review of 12 economic feasibility studies shows that the projected cost of biodiesel from oilseed or animal fats ranges from US$0.30–0.69 per litre, including meal and glycerol credits and the assumption of reduced capital investment costs by having the crushing and/or esterification facility added to an existing grain or tallow facility. Rough projections of the cost of biodiesel from vegetable oil and waste grease are, respectively, US$0.54–0.62 and US$0.34–0.42 per litre. With pre-tax diesel priced at US$0.18 per litre in the USA and US$0.20–0.24 per litre in some European countries, biodiesel is currently not economically viable and consequently its production is largely subsidized.
6. Very few microorganisms are capable of digesting glycerol in an oxygen-free environment. Recently scientists have discovered that *E. coli* is in fact able to metabolize glycerol in a purely fermentative manner by identifying the metabolic processes and conditions that allow a known strain of *E. coli* to convert glycerol into ethanol: S. Shams Yazdania, R. Gonzalez, Anaerobic fermentation

of glycerol: a path to economic viability for the biofuels industry, *Current Opin. Biotechnol.*, 2007, **18**, 213.

7. The spot price or spot rate of a commodity, a security or a currency is the price that is quoted for immediate (spot) settlement (payment and delivery). See also, Oleoline glycerol market report: http://www.oleoline.com/reports.html.

8. L. Ott, M. Bicker and H. Vogel, *Green Chem.*, 2006, **8**, 214.

9. The product (termed NExBTL) is a synthetic diesel fuel, free of oxygen and aromatic compounds. Side-products include propane, petroleum, gasoline, CO and CO_2. The first NExBTL production plant with a rated annual capacity of 170 000 tonnes came on stream in Finland in the summer of 2007. For detailed information, see: http://www.nesteoil.com.

10. D. Nilles, Combating the glycerol glut, *Biodiesel Magazine*, September 2006.

11. K. Shaine Tyson, Biodiesel R&D potential, Montana Biodiesel Workshop (October 8, 2003): leg.mt.gov/content/lepo/2003_2004/subcommittees/energy_group/staffmemos/bio_potential.pdf.

12. G. Graff, *Purchasing* (June 15, 2006), Glycerol glut sends prices plummeting, http://www.purchasing.com/article/CA6341035.html.

13. M. P. D. Heming, *Glycerine Market Report*, No 71, 13 December 2005, http://www.oleoline.com/admin/oleoftp/marketreport/samples/Q_glycerol_sample.pdf. This comprehensive quarterly, published by the glycerol market brokers HBI is recognized as an industry reference by most of the major players in the glycerol market. HBI sold over 150 000 tonnes of glycerol in 2006, more than 10% of the total world market, including an important share in both China and the USA.

14. G. Graff, *Purchasing* (June 14, 2007) Glycerol prices tick up in 2007 as imports slow to a trickle, http://www.purchasing.com/article/CA6450783.html.

15. In February 2007 the US Government Accountability Office (GAO) released a report to members of the US Congress entitled, "Uncertainty about future oil supply makes it important to develop a strategy for addressing a peak and decline in oil production". The report examined 22 studies for estimated timing of peak oil production; 18 forecast peak production before or around 2040, with 2 more by 2050. http://www.gao.gov/new.items/d07283.pdf.

16. A. J. Ragauskas, C. K. Williams, B. H. Davison, G. Britovsek, J. Cairney, C. A. Eckert, W. J. Frederick, Jr. J. P. Hallett, D. J. Leak, C. L. Liotta, J. R. Mielenz, R. Murphy, R. Templer and

T. Tschaplinski, The path forward for biofuels and biomaterials, *Science*, 2006, **311**, 484.

17. R. R. Soares, D. A. Simonetti and J. A. Dumesic, Glycerol as a source for fuels and chemicals by low-temperature catalytic processing, *Angew. Chem. Int. Ed.*, 2006, **45**, 3982.
18. L. Napoleoni, *Rogue Economics*, Seven Stories Press, New York, 2008.

Subject Index

Page numbers in *italics* refer to figures and tables.